Interstellar Travel – An Astronomer's Guide

By
Dr. Sten Odenwald
The Astronomy Café

Praise for Sten Odenwald

"With his first book, *The Astronomy Café*, Sten Odenwald demonstrated that he belongs at the interface between the cosmic frontier and the public inquiry of that frontier."– Neil De Grasse Tyson (American Museum of Natural History).

"Odenwald not only sets a comfortable conversational tone; he adds a sense of humanity that some science books tend to omit" - Sky and Telescope

"Odenwald's concise writing and his sense of humor and layperson's writing style is a laudable public service" – Bloomsbury Review

Other books by this author:

Interstellar Travel: An Astronomer's Guide
Exploring Quantum Space
Ask the Astronomer
Solar Storms

The 23rd Cycle
Patterns in the Void
The Astronomy Café
Back to the Astronomy Café
Stepping Through the Stargate

For related information about astronomy
Visit the Astronomy Café website at
http://www.astronomycafe.net

Copyright © 2015, Sten Odenwald

Acknowledgments

I would very much like to thank the many readers of my Huffington Post Blogs who made so many excellent comments about my 'Interstellar Travel' essays, and who inspired me to write this more-comprehensive guide!

I would like to thank Dr. Jason Kring at Embry Riddle University, and President of the Society of Human Performance in Extreme Environments, for the many discussions we had about humans functioning in space.

Finally, but not last, I would like to thank the many science fiction authors who inspired my imagination and kept me motivated all these years to consider interstellar travel an amazing opportunity for humanity

…if we could only figure out how to do it!

Table of Contents

Prolog 1
The History of Interstellar Travel 3
Parallel Universes 21
Hyperspace 25
The True Shape of Space 47
Where would we go? 59
Habitable Zones 65
The List of Destinations…so far! 71
Flare Stars 85
Are we there yet? 93
Exoplanet Atmospheres 101
A Matter of Extreme Gravity 115
Are we missing something? 119
Planetary Moons 127
Space is not at all empty! 129
Radiation 145
Mutations 163
Diseases 169
Physiological Effects 179
Psychiatric Effects 185

Interstellar Travel

Us versus Them	199
Because it's the Law!	205
What's for dinner?	213
Funny Odors	221
Diurnal cycles	225
Communication	229
Things that Break	245
It's cold in space!	255
Fuel and energy	261
Down and Back	283
How much will it cost?	291
The Miracle Cure: Stasis?	301
Artificial Intelligence and Virtual Reality	307
Interstellar Rocket Technology	319
Magic and Advanced Tech: The Big Lie	343
Where is everyone?	347
The Pessimist's View	353
The Optimist's View	357
Conversations with the Public	361
Epilog	373
Image Credits	376
Bibliography	380

Prolog

Interstellar travel.

Just the words taken together resound with adventure and unimaginable mystery. Since childhood, most of us have been reading about distant worlds and brave empires standing against the incredible vastness of galactic space. We all have our favorite authors who masterfully recount the human challenges of exploring deep space and of course encountering mysterious aliens. We seldom question the underlying science or engineering miracles that make these stories possible. We read the stories for the captivating glimpses they give us of human adventure and galactic mystery, not as physics textbooks. The best authors, however, toe the line with established science and engineering, but base stories on small steps beyond the known.

What distinguishes science fiction stories like *2001: A space odyssey* from fantasy stories like the famous Harry Potter series is that in science fiction, the technology at least sounds plausible, and enables the story to develop. Magic wands and teleportation as the staples to fantasy stories make no scientific sense and SF readers become negatively distracted by their implausibility.

The advent of movies such as the *Star Trek* and *Star Wars* series, along with the recent *Interstellar* has once again stimulated interest in interstellar travel, not just as a well-worn science fiction conveyance, but possibly as a real-world goal we should attempt to reach in the centuries to come. But what does that really mean?

Interstellar Travel

Most of the modern discussions now revolve around the technology needed to make multi-light year journeys, but the focus on propulsion technology is a distraction from truly contemplating the vast distances and challenges that such trips actually entail.

As an astronomer, I can't speak to the finer details of future technology, but I can try to define what such journeys are likely to be in general terms. Issues like fuel, time, energy, shielding, human behavior, artificial intelligence are most certainly just as important as the technology that gets us to the stars. Will hibernation work? How would we replace things when they break? Will our immune systems completely collapse during the journey?

And then there are the astronomical issues...My specialty!

Where would we go? Will we really spend trillions of dollars to just 'point and shoot' and randomly select a destination? When we get there, will we be content to live under a dome as a colonist? How do we deal with the hazard of interstellar clutter encountered at flight-speed?

If you are interested in 'sneaking up' on interstellar travel by exploring these issues, this book is for you. We will define what such journeys will look like by looking at small pieces of the puzzle and seeing how they restrict the kinds of solutions we can contemplate. You will be amazed how a little math will shed new light on many issues we take for granted in planning the last great expansion of humans into the cosmos!

The History of Interstellar Travel

Imagination is the engine that propels all of our many creative endeavors. It is a gift from an organ that compulsively asks 'How?', 'Why?', 'Where?' and 'What if..?' The human brain is relentlessly inventive. It constantly searches for patterns from the data it receives. Lacking data from the outside world, it will patiently bide its time by creating its own reality stimulation. Nowhere is this insatiable pattern searching more in evidence, and more volatile, than in the roller coaster journeys we take while asleep. Our waking state also benefits from this activity by sensitizing us to possibilities that we might not have anticipated. The structure of the benzene molecule was the product of a dream by Chemist August Kekule in which he saw a snake biting its tail to form a ring or carbon atoms. The principles of relativity were discovered when Albert Einstein imagined what the world would look like if you could ride on a light beam. The concept of the 'bubble chamber' was developed by Donald Glaser while watching the suds form in a beer stein. These examples don't mean that unrestrained imagination always results in new ideas in physics, but surely a playful approach to understanding nature, together with the right technical background, can work wonders from time to time!

Science fiction authors have for decades created universes in which a bewildering variety of answers have been proposed to the dilemma of traveling to the distant stars and beyond. Surprisingly, these answers have followed an almost evolutionary sequence that runs nearly parallel to developments in 20th century physics. Of course, some authors have been more in touch than others with the reality and limitations of the

Interstellar Travel

physical world, but then again, we all understand that science fiction is only meant to be a plausible image of a future world. It is not supposed to be a slavishly literal extension of our present scientific knowledge. Of greater consequence to nearly all SF stories is how humans heroically resolve the old conflicts of the human 'condition' but in the context of new environments. The environments often mirrored the political conflicts of their times with fierce, interplanetary battles dominating the SF of the 1930's and 40's.

It is easy to understand how SF got such a bad literary reputation in reading some of the stories that made it into print during the early 1940's. Hostile aliens from Mars and Venus fought interminable battles with earthlings. Earthlings traveled interplanetary space in ships shaped like missiles, footballs or dirigibles, powered by 'rocket tubes'. Nearly all the planets in the solar system were occupied by implacable and often belligerent aliens. Occasionally aliens would enter our solar system from distant worlds with colorful names like 'Boron' or 'Talpite'. Stories were often shallow, with little character development, serving only as frameworks for presenting some bizarre technical gimmick.

The History of Interstellar Travel

The physical principles underlying science fiction writing during this period were a complicated pastiche of ideas derived at least in spirit from the sweeping developments then occurring in relativity theory and atomic physics made during this time. Most of these developments were only partly understood by the authors of that time, which is probably why SF stories often hinged on outright errors even in some of the most elementary aspects of the physical world. For instance, the Bohr atom in vogue during the early 20's with its planetary electrons was literally a microcosmic solar system. Our solar system was in turn simply an atom in a much larger universe. Several authors imagined people taking journeys to these other worlds within the atom by using machines that either shrank or enlarged their bodies by suitable scales as in S.P. Meek's *Submicroscopic* (1931), or G. Peyton's short story *The Man from the Atom* (1926). In the later instance, probably the earliest story of this kind, one Prof. Martyn builds a machine that subtracts or adds atoms to the human body until it has grown or shrunk to the desired size.

Between 1924 and 1927, the physicists Heisenberg, Schrödinger and Paul A.M. Dirac had all but written the last pages of the modern theory of atomic structure based on quantum mechanics. The Bohr-Sommerfeld 'planetary' atom though conceptually simple was, nevertheless, invalid so that there was no longer a basis for thinking that electrons were miniature planets with solid surfaces by this time. Yet SF based on this curious principle of 'worlds within worlds' persisted even as late as 1949 in Stan Raycraft's *Pillars of Delight* and in *He Who Shrank*.

With the establishment of the principles of quantum mechanics in the 1930s and 40s, there came a minor resurgence of interest

Interstellar Travel

in the tantalizing properties of the atomic world. In James Blish's story *Nor Iron Bars* (1956). By some unknown, accidental means, the 'Haertel Overdrive' endowed the ship Flyway II with negative mass. The ship was promptly ejected from normal space and took up residence within an atom. Presumably, this was the only place in the universe where negative mass could exist. In many ways this idea resembles the old Aristotelian notion that matter has an innate sense of its proper place in the universe, and like a stone falling to the ground, will seek out its natural resting-place. The ship could not logically exist in our universe, so it found a place where it could. The captain also sets foot on the surface of an electron described as "... *a swirling, opalescent substance...covered with fine detail and rills rather like mercury...Its boundary trailing off into space indistinctly [since] the electron never knows exactly where it is...*" The electron was clearly assumed to have a finite size, though in deference to quantum mechanics, there was a haziness about its boundary.

The passengers of Flyway II, meanwhile, were having difficulties of their own, experiencing telepathic effects also forbidden in the real universe. The author raises the possibility that telepathy and quantum mechanics might be related in some unknown manner. This idea that psychic phenomena are in some way a consequence of the quantum world also appears in Colin Kapp's *Lambda I* (1962). In this short story, a passenger ship travels directly through the Earth in something called Tau-space. The idea is that if you vibrate a mass just right, it can be made to pass directly through any matter including the entire earth. Unavoidably because of the density differences in the strata that it passes through, upon arrival at its destination, a tauship re-materializes in some harmonic vibration of the proper mode and

The History of Interstellar Travel

must be 'kicked' into the correct oscillatory mode for recovery. Evidently, mode slips can also be triggered by extreme emotional or psychic states, and like Lambda I, cause the ship arrive in the unrecoverable 'omega' state. An earlier short story *The World Beyond* by Guy Archette in 1947 describes the principle behind this vibratory model for matter as "*Everything is made up of atoms, and there are spaces within the atom fully as vast as those between the planets in the solar system. The spaces ... may be occupied by the components of a hundred other atoms, each possessing a different vibration rate, and each vibration rate constitutes another world.*" It is difficult to believe that Colin Kapp and Guy Archette were not drinking from the same tap in inventing this unusual world view!

James Blish's second attempt at utilizing quantum effects to drive spacecraft appears in his epic novel *Cities in Flight* (1958) based on a collection of novellas written between 1950 and 1957. As the story unfolds, we learn that during the first decades of the 21st century, western scientists had discovered the 'Dillon-Wagoner gravitron polarity generator' which immediately became known to the engineers as a 'spindizzy'.

The basic operating principle as described in *Earthman Come Home* (1950) is that all rotating bodies produce magnetic fields whose strength is proportional to their rate of spin, their mass, and the constant of gravity. What a spindizzy does is to alter the magnetic moment of every atom within its field thereby changing the constant of gravity. *Cities in Flight* is one of the few stories I've read that actually gives an equation, identified as the fictitious Blackett-Dirac Equation, claiming to show how 'G' can be altered by changing a particle's spin, to wit : $G^2 = 8 P c / U$ where P is the magnetic moment of the body, c is the speed of

Interstellar Travel

light, and U is the angular momentum! Evidently, by increasing U for every particle in a body, the constant of gravity is reduced, ergo the name spindizzy. A space drive based on this principle, for reasons not made clear by the author, has no practical upper limit to its speed, and can break the light speed barrier by drawing less than a few watts of power. Entire Earth cities were soon equipped with spindizzies and set-out to colonize the galaxy. The story contains several amusing anachronisms. For example, although much of the action of the novel takes place between 3000 and 4000 AD, the engineers are still using slide rules!

Another story mentioning an association between magnetic phenomena and gravity is found in George O. Smith's *Meddler's Moon* (1947). The so-called 'Hedgerly Effect' proved that there is a relationship between magnetism and gravity which led to the electromagnetic control of gravity. Artificial control of gravity fields allows the gravitational mass of any particle to be altered at will. This does away with rockets that have to throw lots of reaction mass out their tails, since mass approaches infinity near the speed of light. Since mass can now be controlled, it can be

The History of Interstellar Travel

reduced so that travel near the speed of light is possible.

Although interplanetary travel by conventional rockets was the mainstay of most SF during the 1930's and 40's, traveling through space to the distant stars has long been known to entail lengthy journeys and posed a whole host of other problems: Problems that were generally so challenging that SF authors tended to stay away from interstellar travel during much of this time. The most straightforward way to shorten the travel time is simply to increase the velocity of your rocket. Sometimes, exotic new rocket fuels were invoked like 'mercuron' in Robert Willes' *Orbit XXIII-H* (1938), capable of producing exhaust velocities of 65 km/sec. For long journeys to Mars and beyond, the crew would take various combinations of sleeping drugs.

With the coming of the Atomic Age these pure, brute force methods were upgraded by invoking the mysterious *deus ex machina* of 'atomic motors'. Velocities up to and including light speed could now be reached as in Richard Tooker's *Moon of Arcturus* (1935) but without the travelers experiencing any time dilation effects. The spaceship Meteor III powered by the energy of "disrupting carbon atoms" takes 26 years to travel to Arcturus with its 18-man crew. This is a remarkable story since much of the work on atomic energy, specifically the fission mechanism, was not readily accessible to the popular press. The mysterious concept of 'atomic drive' also appears in A.E van Vogt's 1940 classic *Slan*, but is not described other than to note that it is based on anti-gravity propulsion. Anti-gravity is an elusive principle in SF; a catch-all for achieving nearly unlimited velocity by mysteriously canceling the force of gravity. Murray Leinster's *First Contact* (1945) also refers to a spaceship which

Interstellar Travel

can travel at 'speeds incredible multiples of the speed of light' to take pictures of a supernova explosion.

In general, rockets blasted off from Earth under constant accelerations of 1- 2 gravities; the maximum that humans could comfortably endure for long periods of time. The arithmetic of constant acceleration is compelling when applied to space travel.

At '1 G' boost, the travelers feel exactly like they are standing on earth, yet the ship will achieve 50 percent of the speed of light (50%C) in about 6 months after having traveled 0.1 light year! A trip to Mars at this acceleration would take about one month with a terminal velocity of 10%C. If you were willing to endure 2-Gs, you would get to Mars in 10 days with a terminal velocity of 6%C. Inertialess drives, as they came to be known, were often introduced to achieve even higher accelerations. If you

The History of Interstellar Travel

had an 'acceleration compensator' that allowed the ship to move at 1000-Gs but canceled all but 1-G for the humans, even faster trips could be made.

For many years 'souped-up' rockets violated natures prohibition on material bodies traveling faster-than-light (FTL). Some stories took great relish in how badly they could break this speed limit. Edward E. Smith's *Grey Lensman*(1939) describes how spaceships travel at speeds of 60 parsecs/hour (2 million times c) in interstellar space and up to 100,000 parsecs/hour (3 billion times c) in the more rarefied intergalactic space by using 'cosmic energy'. These ships used the inertialess drive developed by Bergenhlom and which opened up commerce throughout the Galaxy. The sequel to this story *Second Stage Lensman* also mentions a race called the Medonians who installed an 'inertia-neutralizer' on their home world and moved it to Lundmark's Nebula, also known as The Second Galaxy. Robert Heinlein's *Methusala's Children* (1941) describes a "spacedrive that uses light pressure under conditions of no inertia to travel just under the speed of light.

Other mechanisms were also advocated such as the one found in Rog Phillip's short story *Starship from Venus* (1948). A mysterious starship from Venus lands on Earth and earthlings soon learn the secret of interplanetary travel. We learn that electrons and protons have opposite inertia (they don't). By shooting protons out of one end of the ship and electrons out the other, the net impulse is magically in the forward direction at 1/3 the velocity of light.

Interstellar Travel

The 1965 novel *Rogue Ship* by A.E van Vogt tells the tale of the 'Hope of Man' and its crew who were boosted to nearly the speed of light in order to take advantage of the time dilation effect promised by the 'Lorentz-Fitzgerald Contraction Theory'. The ship was supposed to eject reaction mass at near light speed so that it gained mass, which multiplied the effectiveness of the reaction mass so that *"...a thimble full of mass could give almost infinite reaction power"*. The expected effect never materialized and the Hope of Man wound-up taking decades to reach Alpha Centauri. There was also the claim that at the speed of light, mass becomes infinite but the volume of a particle vanishes so that matter ceases to be subject to inertia. Both claims are of course invalid, but the stories themselves are entertaining. Another story that bases its propulsion on an imaginary twist to relativity theory is Ross Rocklynne's *The Moth* (1939). A 'reverse contraction' device shrinks a ship by collapsing electron orbits, this is necessary because *"...if you decrease the length of a ship to zero, it automatically assumes the speed of light..."*. To gain any speed without acceleration, *"...you just shorten its length commensurate with the speed you want."*

Poul Anderson's novel *To Outlive Eternity* (1967) followed the edicts of special relativity with some fidelity, and explored a very important limitation to relativistic travel: where do you get the reaction mass to continuously accelerate the ship? The answer is that instead of bringing it with you, you scoop it up from the interstellar medium as you go!

This story followed the exploits of a hapless crew in an interstellar ramjet ship whose deceleration mechanism had failed. Their only solution was to drive the ship ever faster to

The History of Interstellar Travel

within a stone's throw of the speed of light. At these speeds the time dilation effect would allow them to outlive the collapse of the universe. They continuously picked up speed by first passing through dense interstellar clouds in the Milky Way, then upon leaving the galaxy, heading for other galaxies, which they would pass through in a matter of minutes. Eventually, they managed to avoid the re-collapse of the universe. Artistic license enters the story with a vengeance when, during the universe's re-expansion at nearly the speed of light, the crew coasted along with developing galaxies and solar systems at nearly zero relative velocity, until they found a suitable Earth-like planet to colonize.

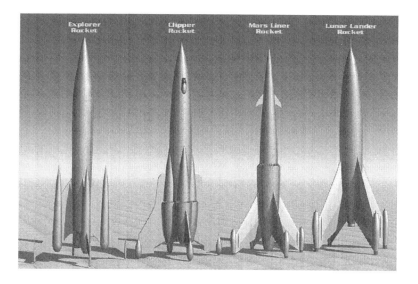

The first test flight of an FTL ship using a hydrogen drive is described in Walt Sheldon's 1950 short story *The Eyes are Watching* in which no one really knew what would happen when a ship surpassed the speed of light. Some of the scientists are acknowledged to have their doubts that this velocity barrier

Interstellar Travel

could be broken, and even speculate that " you might warp over into another dimension". An earlier story by George O. Smith *Pattern for Conquest* (1946) refers to something called the 'superdrive' which allows space ships to accelerate to very nearly the speed of light, but understandably enough, the details are not presented. In the same story, we also hear of 'tractor' and 'pressor' beams which can be used to 'tear the guts' out of an enemy ship. This technology is based on something called the 'space constant adaptor' and is described in Murray Leinster's *Adaptor* and in *The Ethical Equations* (1945). An earlier story *Redevelopment* written by Wesley Long and published in 1944 describes a round trip to Sirius in 6 months using 'gravitic generators' and particles called alphons, which are used to propel the ship past the light speed barrier. This mechanism, by the way, was also termed 'superdrive'. P. Schuyler Miller's short story *Gleeps* (1943) refers to 'warpships' which are used in interstellar and possibly inter-universe travel. Generally, the development of FTL technology is seen as prohibitively difficult. Usually, humans stumble upon its secrets accidentally, or a super-civilization gives this knowledge to us. This generosity is not always without major cost.

Rendevois in Space by Guy Archett in 1949 described the discovery of an alien spaceship beyond the orbit of Saturn capable of interstellar travel. But the aliens who own it refuse to tell the earthlings the secret of how it works. Earthlings are still too war-like to be allowed to learn of such a powerful technology. This notion that giving earthlings the secret of interstellar travel would be disastrous for galactic civilization is also found in Robert Moore Williams *Star Base X* (1944). In this story, aliens known as Ahrneds (rhymes with 'airheads'?) refuse to give up

The History of Interstellar Travel

this secret because they realize the inevitability of interstellar war once the aggressive and competitive earthlings get out among the stars. *Homo Sol* (1940) by Isaac Asimov also describes how a galactic federation of civilizations invited Earth to join after humans had discovered the secret of interstellar travel and arrived at Alpha Centauri with the intent of colonizing its fifth planet. Human inventiveness actually had outdone the rest of the galaxy by developing an improvement on 'hyperatomic' drive that was superior to anything that the Federation had. Operating principle: Unknown. In a short time, humans had also managed to transform many benign devices used by the Federation into astonishingly lethal weapons!

Then came a fascination with the properties of anti-matter. John Bridger's *I'm a Stranger Here Myself* (1950) refers to a method for FTL travel called "multi-phase travel" which is based on transforming terrene matter into contra-terrene matter, what we now call simply matter and anti-matter. For some reason, anti-matter is claimed to travel faster than light. But again, this is not a property that earthlings discovered, instead they learned it from a benevolent galactic super-civilization whose emissaries visited us. Then again, even some super civilizations are not omniscient.

Michael McCollum's *Life Probe* (1983) and *Procyon's Promise* (1985), for example, have Earth visited by a ship from a very old civilization called The Makers. After millions of years the Makers had given up trying to develop a FTL drive, even though they had developed several independent theories that showed FTL drives were, nevertheless, physically possible to build. They turned to making contact with other civilizations in the galaxy

Interstellar Travel

that might have stumbled on the right engineering ideas. Thousands of automated 'slow boats' driven by fusion engines, powered by 'I-mass' Hawking Singularities, and traveling at sub-light speed, were dispatched into the galaxy in search of more cleverer civilizations to tell them the secret of FTL travel. One of these probes wound up in our solar system and becomes the center of, you guessed it, interplanetary intrigue. Eventually, earthlings take up this quest and find the pieces to a derelict FTL ship in the Procyon system. They discover that the Makers had already learned the secret to FTL travel centuries before the Probe entered the solar system, and had abandoned their home world. This idea that FTL travel is already known to some other civilization can also be found in *Nomad* written by Wesley Long in 1944.

There may also be a technical problem with developing FTL travel. Asimov's *Paradoxical Escape* (1945) describes the search for the secrets of interstellar travel in which a mechanical brain is fed everything we know about astronomy, physics and something called 'space warp theory'. The 'Brain' eventually figures out the secret, unfortunately such trips would be fatal to humans. This is why previous attempts by other mechanical brains had failed. Since a robot cannot create anything harmful to humans, previous robots literally fried themselves rather than break this 'First Law of Robotics'.

A new entry into propulsion technology appeared in 1986 and is found in Arthur C. Clarke's *The Songs of the Distant Earth*. Once again, FTL travel is assumed to be impossible, and is flatly disavowed by Clarke in the books preface as fantasy not science fiction. However, a virtually unlimited energy supply for sub-

The History of Interstellar Travel

light travel is tapped in the quantum fluctuation of space at the Planck scale. In the 36th century, the ship Magellan draws its boost energy directly from the energy of empty space so that it no longer needs to carry reaction mass with it. Interstellar travel faster than 20%C is acknowledged to be dangerous because the stray atoms the ship encounters act like miniature hydrogen bombs as they strike the ship. Each ship is equipped with an ablative 100 kiloton cap of ice on its forward edge. Conveniently, not even the ship's captain or its crew members really know how the quantum drive works, but such delightfully cryptic explanations as "*fluctuations in the geometrodynamical structure of 11 dimensional superspace*" can be found in the novel. A similar tale of tapping the energies inherent in space is found in William Lawrence Hamilton's *Planet of Duplicates* (1945). This energy is in the more prosaic form of the exhalations of matter and energy from all the stars in the Milky Way.

Several stories such as Gordon Dickson's *Mission to Universe* (1965) and Poul Anderson's *Door to Anywhere* (1966) are hard to classify since they attempt to use the vagaries of quantum mechanics and antiquated cosmologies to find shortcuts through space involving 'Phase shifting' or a modification of Hoyle's Steady State cosmology in 'jumpgates'. Jumpgate technology as explained in *Door to Anywhere* is highly suspect. A similar access way to distant worlds can also be found in Oliver Saari's *The Door* (1941) in which a gateway is found among the ruins of an ancient city in the Sahara Desert. Stepping through, the traveler is shifted to the surface of a planet in a binary star system. Unfortunately, the origin and function of this gate are not as important to the story as the search for it.

Interstellar Travel

Phase shifting is described in *Mission to Universe* at least in interesting terms that sound plausible if you don't think about them to deeply. Heisenberg's uncertainty principle states you can never know both a particles position and velocity with absolute precision. In phase shifting, the velocity of a ship is measured precisely so that the ships wave function spreads out over interstellar distances. By some magical process, the ship's wave function is encouraged to 'peak up' at some target position several light years distant. When the wave function collapses, the ship then shifts and takes up residence at the target position without having moved through space. The implication is that the mere knowledge of the ships precise velocity as determined by the shipboard computer, is enough to change its wave function, much like the act of observing an electron automatically places the electron in a definite state, were before it could have been anywhere.

Among the technological innovations by the super race known only as the Puppet Masters were 'stepping disks' and 'transfer booths' as described in Larry Niven's 1970 novel *Ringworld*. These were, evidently, teleportation devices of some kind. The technology of matter transmission is also used and described in the stories by Alexander Blade such as *The Vanishing Spaceman* (1947).

Clifford Simak's *Way Station* (1964) utilizes a galaxy-wide network of 'transfer booths' also developed by a mysterious super-civilization. The operating principles are not described, but resemble teleportation (materialization) chambers. The extensive references to 'telepathic' aliens and the manipulation of aspects of the physical world entirely unsuspected by humans,

The History of Interstellar Travel

resembles magic in all but direct citation. George O. Smith in *Special Delivery* (1945) describes how matter transmitters scan matter atom by atom and then disassembled the body, storing its raw atoms in a 'matter bank'. Matter itself isn't transmitted, but the information and energy released in the disassembly is beamed to a second station, which uses the raw materials in its matter bank to re-create the body atom by atom. A.E. van Vogt's *The Mixed Men* (1945) also refers to matter transmission, this time in two distinct modes. Earth is the center of a 3 billion solar system empire where ultrawave radio provides instantaneous communication. People move about either by electronic image transmission followed by reconstruction from local organic material, or conversion of the body into a flow of electrons transmitted through space and then rebuilt at the destination.

An interesting twist on teleportation devices is found in Robert Abernathy's *The Canal Builders* (1945). Although 'teleports' are the standard means for moving around in the solar system, a thrill-seeking earthling builds a space ship to travel through space to Mars the old-fashioned way, just to be unconventional. The trip takes two weeks, and upon arriving he discovers the ruins of a long dead civilization at the spot where an Earth city should have stood. It turns out that in the 'interspace' in which teleportation operates, there is also a time shift involved. Rocket travel lands you on Mars in 'now plus two weeks' but teleportation takes you to Mars in 'now minus 2000 years'. The ruins were those of an Earth colony built 2000 years ago! Since matter-transmission and teleportation are more the stuff of ESP and ghost stories with no obvious physical mechanism or

Interstellar Travel

science behind it, I'm not going to dwell on this 'technology' further.

By the 1960's some calculations by physicists had turned up odd particles called tachyons whose slowest speeds were that of light. Most physicists just assumed these were fictitious particles that indicated a theory had broken down, but science fiction authors weren't so sure. In *The Sins of our Fathers* (1976) written by Stanley Schmidt something called a para-tachyonic drive can boost the ship to any velocity faster than light without expending much energy, but travel within a few percent of the speed of light using this 'Rao-Chang Drive' requires the expenditure of enormous quantities of energy.

No one knows how the drive works because the operating principle was discovered accidentally. The younger generations of physicists are, furthermore, not interested in overhauling physics in order to accommodate it! It is not clear from the SF technology in the story whether the ship is converted into 'tachyon matter' in ordinary space, or whether the ship enters some other universe continuum where tachyon physics is valid.

Parallel Universes

Although faster-than-light travel might be impossible, and travel to other stars difficult or out of the question, some authors over the years have proposed staying right where we are in space. Time travel or travel to parallel universes have in recent years become viable alternatives to what has always appeared to be the inevitability of interstellar travel. These modes of travel violate no principles in relativity since they take place in a completely different arena than ordinary space-time.

Time travel based on 'the fifth to tenth dimensions' is referred to in *Rescue into the Past* (1940) by Ralph Milne Farley. From time to time, there have also appeared a variety of even more spectacular stories that aggressively push at the very meaning of Reality. As early as 1915 in *A Drop in Infinity* written by Gerald Grogan, a scientist named Hubble-Bubble creates a machine based on electricity that projects a person into the 4th dimension. According to Dr. Hubble-Bubble, certain waves run along the 4th dimension which is somehow counteracted by other waves that run the other way, and that we perceive as electricity. By tuning the chamber, the electrical field inside lets you experience these other dimensions. The traveler can then visit what can only be called parallel Earths. These worlds mysteriously coexist with ours, spatially like the rungs on a single, but infinite, ladder. This idea that the proper application of electrical fields might open a magical, inter-dimensional gateway re-appears almost word for word in Charles Recour's 1949 short story *The Swordsman of Pira*, in which we hear that "*if an object were suddenly thrust into a strong electric field, it would be rotated*

Interstellar Travel

through a warped space into a fourth dimension" Space is warped by strong electrical fields so that a human can use this to travel into other parallel 'time streams'.

This parallel-Earth idea is also explored in George Scheer Jr's *Another Dimension* (1935) where a spaceship sent to Mars is caught in a 'whorl in space' that hurles the ship into another dimension. There the explorers discover another Earth and solar system at the same location as ours, but not at all like the Earth they left behind.

Interesting stories have also been developed around the idea that, although FTL corporeal travel may be prohibited, the psyche may travel at whatever velocity it chooses. Certain mixtures of psychotropic drugs such as morphine or belladonna could send a voyager's psyche into higher dimensions as in Stanton Coblentz's *Beyond the Universe* (1934). This story describes a journey into a higher dimension of space that seemed to have taken years and centuries, but in fact the traveler had actually been in a coma for 6 months. During his journey he watched as the universe shrank to a point and directly experienced the *'nothingness at the boundary to space'*. A similar story of psychic travel occurs in Charles Eric Maine's *Timeliner* (1955), in which the psyche of a physicist is propelled our of normal space-time into a 'hyperspace' where time becomes one of the spatial dimensions. In a process termed 'psycho-temporal parasitism' he time travels into the future entering the mind of one human after another.

It has also been proposed that it may be easier to get massless particles like photons, neutrinos or 'exotics' to give us images or

Parallel Universes

messages from other times and worlds. Donald Bern's *Three Wise Men of Space* (1940) described beings who communicate with rays that travel faster than light since they possess a knowledge far beyond that of Einstein. Light rays are used in Richard O'Lewis' *Adam's Eve* to obtain images through the fourth dimension of past ages, including an image of Eve starving in the Garden of Eden! Edmund Hamilton in his short story *The Star Kings* (1947) describes earthlings from 200,000 years in the future who communicate with 20th century earthlings via thought waves, which are the only things that can pass through time because they are not material objects. Interstellar travel in this future age is by means of 'sub-spectrum' rays of the 'minus 30th octave' which are produced in atomic turbines. Also in that age, the relativistic increase of mass with velocity has been circumvented so that 'star-ships' equipped with artificial mass (inertia) control can travel 1000's of times the velocity of light. A recent entry into this genera of using massless particles to glimpse the future is James P. Hogan's *Thrice Upon a Time* (1980).

To Follow Knowledge (1942) by Frank Belknap Long portrays the universe as a superposition of time frames; one for each body in motion. If you moved at the speed of light, everything stands still, but at FTL speeds even time begins to undo itself. All events take place in a 5 or 6-dimensional time track that connects the 'nows' of every frame of reference moving in space. Even the use of mental powers to 'warp space-time' and return to the past has been considered, and can be read about in the 1943 story *Shock* by Lewis Padgett.

So you see, science fiction is filled with clever ideas about faster than light travel. If physics and understanding the cosmos were

Interstellar Travel

only a matter of cobbling the right words together, we should have been building these kinds of ships buy now. But the frustrating thing is that we have gained considerable experience in the last 100 years boosting matter to 99.999...% the speed of light. In no experiment devised so far have we found anything in the distant decimal points of the data that defeats Einstein's special relativity. The injunction that the speed of light is the maximum speed limit for energy and information, or any other 'thing' that has zero rest mass, seems locked-in to our universe.

Hyperspace

Along-side these stories that more or less toe the line with special relativity by proclaiming that FTL travel is impossible, are stories that took the next step following the developments in Einstein's General Theory of Relativity. Authors began, tentatively at first and then with increasing boldness, to find technological solutions to space travel that did not involve moving through ordinary 3-dimensional space with its nasty "thou shalt not exceed the speed of light" edict. These methods might be termed 'inter dimensional travel' since that is often the manner in which the problem is solved. This sub-genera of writing probably had its inception when authors began exploring certain conceptual ideas in general relativity. As they became more comfortable with the ideas of multi-dimensional space, authors developed an increasing array of applications for it. At first, these journeys were limited to laboratory experiments that went badly wrong.

The story by Donald Wandrei *The Blinding Shadows* (1934) described an inventor who builds a complicated optical machine that rotates in all manner of directions, with lenses made of rhillium. This fictitious element has the unusual property, according to the stories Dr. Dowdson, that 50 percent of the energy falling on it just vanishes, presumably into the fourth dimension. The rotation of the rhillium lenses are so complex that, not only does motion occur in the normal 3 dimensions to space, but occasionally into a fourth as well. The good doctor's goal is to image objects from the 4th dimension in 3-dimensional space. What happens is that the black shadows of

Interstellar Travel

alien beings soon appear, escape from his lab, and start to gobble up most of the inhabitants of New York City!

This idea that some new element might emit or transform radiation in this way also appears in Milton Smith's *The Mystery of Element 117* (1949). Our universe extends a short distance into a fourth spatial dimension. Because of this, it is possible to rotate matter completely out of 3-space by building a '4-dimensional translator'. Element 117 is a fictitious magnetic monopole substance, which can be made into an optical lens and used to open a window into this new dimension. It turns out that this other world is the one inhabited by already dead humans. They live in a neighboring world to ours, but just slightly shifted from ours along the fourth dimension. We also read in this story that our 3-dimensional space is but one 'hyperplane of hyperspace'. Succeeding layers are linked together via their quantity of imaginary mass just like page numbers in a book.

Rog Phillips in *The Cube Root of Conquest* (1948) proposes that we co-exist along with other universes in space, but are separated in time. These universes are separated from one another along a 3-dimensional time continuum that exists in the 'imaginary' direction from normal 3-dimensional space. Travel to these parallel worlds requires solving a cubic equation, whose roots give the proper time-like shift to enter these worlds.

In 1940 Robert Heinlein's short story '*...and He Built a Crooked House*' followed the misadventures of a California architect who built his house to resemble the projection in 3-dimensional space of a 4-dimensional hypercube; a shape identified in the story as a 'tesseract'. An earthquake triggered the collapse of this

Hyperspace

inherently unstable shape into a real hypercube with amusing consequences for its inhabitants! Some windows, for example, look out over a desert or a street intersection from a vantage point hundreds of floors above the ground. Another window even looks into Nothing. This 'nothing' is described as a view of a place where space doesn't exist at all, lacking color, form or magnitude. Evidently pure nothingness is an unsettling thing for the human brain to try to interpret, giving the residents of the house a very unsettling feeling. 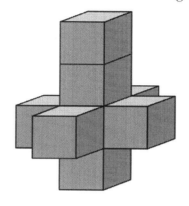 The idea that entry into other dimensions can be caused by sufficiently violent natural phenomena also appeared in *Orphan of Atlans* (1947) by William Hamling. A natural cataclysm unleashes forces and *"...a rent was made in the ether itself....A great space warp was formed around [Atlantas]"*. This catapulted the last few survivors of Atlantis out of their normal space-time and into the future, to arrive in the 20th century.

Alan E. Nourse's *Tiger by the Tail* (1951) describes a civilization living in the 4th space dimension that manages to coerce a human shoplifter to send them more aluminum through a gateway into our dimension that resembles a pocketbook. The shoplifter is apprehended by police who divine the purpose of the pocketbook immediately. Lowering a hook into the pocketbook, they manage to *"pull a non-free section of their universe through the purse, putting a terrific strain on [the alien's] whole geometric*

Interstellar Travel

pattern" a fact which the humans now use as a ransom against invasion.

A similar story of human misadventure in the 4th spatial dimension can be found in Arthur C. Clark's short story *Technical Error* (1950). A technician working in the bowels of a superconducting electric generator was rotated through the 4th dimension due to an unexpected power surge in the magnetic field. He becomes laterally reversed and can no longer metabolize food unless it is provided to him in the 'left-handed' state. As the story goes, the magnetic surge produced a momentary extension into the 4th dimension termed 'hyperspace' by Dr. Hughes. Since time is the 4th dimension, the doctor reasons that the actual rotation must have been through the 5th dimension. The Doctor explains that *"...space of several million dimensions has been frequently postulated in sub atomic physics"*. Of course, no such statement of this kind ever appears in real scientific literature by 1950.

In Clifford Simak's *Shadow of Life* (1943) we hear of Martians who had learned how to shrink themselves to subatomic size by extending themselves into the fourth dimension, causing them to lose mass and size in the other three dimensions. That all matter has some extension into higher dimensions is also stated in *Simultaneous Worlds* (1938) by Nat Schachner. A machine is used to image the supposed 'heavy photon' precursors to cosmic rays. The images formed turn out to look a lot like Earth, but with subtle variations. The idea is soon developed that, since all matter has wavelike properties, more than three dimensions are required to describe matter. Every particle extends into higher or 'ultra' dimensions, which can be imaged using this device,

Hyperspace

however, there is also an inevitable time displacement between these alternate Earths.

The first usage of the term 'hyperspace' is difficult to track down, but by 1950, readers of the magazines <u>Amazing Stories</u> and <u>Astounding SF</u> had already been introduced to it several times. By this year, stories such as Robert Abernathy's *The Ultimate Peril* describe Venusian psycho-physicists attacking Earth with hyperspace weapons, and S. M. Tenneshaw's *Who's that Knocking at my Door?*, about a honeymooning couple whose hyperdrive breaks down near a white dwarf star en route to Deneb. The origin of the technology becomes, to some authors, an impossibility for Earth scientists to have figured out by themselves.

Secret of the *Yellow Crystal* (1948) by Guy Archette, a mysterious crystal is found on Mars among the empty ruins of the Martian civilization. The Thulani as they were called, knew how to rearrange the molecular structure of crystals without mechanical technology, to tap 'extra-dimensional or sub-spatial energies'. They also knew about hyperspace, and apparently used it in some unfathomable way to leave Mars rather than face extinction. This notion that, just as for FTL travel, humans had to be shown how to use hyperspace or 'space warps' for spaceship propulsion also appears in Nelson Bond's 1943 sort story *That Worlds May Live*.

We also hear of "warp generators" in *The Flight of the Starling* by Chester S. Geier in 1948. The maiden flight of the research vessel 'Starling' involves a circumnavigation of the solar system in 3 hours at a velocity close to the speed of light using atomic-

Interstellar Travel

powered warp generators. These generators *"...create a warp in space around the ship...a moving ripple in the fabric of space."* The ship rides this ripple like a surfboard. The speed of light is acknowledged to be the absolute maximum velocity, however, the Starling is thrown out of normal space and into negative space, whose entry occurs once the ship nears the speed of light. Upon exiting they find themselves thousands of years in the future orbiting another, older Earth. A modification to the 'Hyperspace Equations' showed that in-between normal space and negative space is a zone called hyperspace. They had overshot hyperspace and entered negative space where time travel is possible. To travel to distant points in their own universe, they have to carefully accelerate into the hyperspace.

In 1947, Asimov's short story *Little Lost Robot* has 'Hyperatomic Drive' shortened to 'Hyperdrive' and goes on to describe how *"...fooling around with hyper-space isn't fun. We run the risk of blowing a hole in normal space-time fabric and dropping right out of the universe"*.

The term 'Hyperspacial Drive' also appears in Chester S. Geier's 1944 story *Environment* but aside from the comment that *"... You go in here, and you come out there..."* and that where you come out is uncertain by several million miles, that's all that is said about it.

Nelson Bond in 1943 describes the first artificial space warp into the 4th dimension, but humans are not the ones to have discovered its secrets. Humans have to travel to Jupiter to consult with the scientists there who then show the humans how to build FTL ships. The operating principle of this 'quadridimensional drive' is described as *"...the Jovians create a 4 dimensional space warp between points in 3 dimensional space. A*

Hyperspace

magnetized flux field warps 3 dimensional space in the direction of travel...its as easy as that." Also in 1943 A.E. van Vogt's *M 33 in Andromeda* recounts the exploits of the expedition ship Space Beagle, which receives mental messages from an advanced civilization in the Andromeda galaxy. Earthlings use hyperspace in planet to planet matter transmission. Hyperspace s described as not 'strictly an energy field' but requires external pressure in the form of gas pressure at both the outlet and inlet positions otherwise, the hyperspace opening takes millions of years to heal itself and an explosion could result. Focusing a hyperspace transmitter on a spaceship moving FTL requires specifying coordinates in a 900,000-dimensional space and is impossible to control.

The famous *Lensman* and *Skylark* series written in 1928 by E.E 'Doc' Smith represented a complex universe where some attempt was made to create new physics apparently patterned after field theory and quantum physics developed during the 1930's. Much of the language used to describe the propulsion mechanisms involve terms normally found in nuclear physics such as 'fields', 'rays' etc.

Grey Lensman published in December, 1939 has several references to 'Dirac Holes and negative energy weapons' and we also hear of a scientist who had developed a new math capable of handling 'the positron and the negative energy levels'. The '5th order drive' developed by a scientific race called the Norlamins, could create controlled time warps and allow the travelers to voyage anywhere in the universe at millions of times the speed of light. The only catch is that the rays that give rise to the 5th-order drive are only emitted by a rare element called

Interstellar Travel

rovolon: an element found only in some stars. To get to these stars, which can be identified spectroscopically, you have to travel for years at sub-light speed. Also in *Grey Lensman* the Boskonians attack the Lensman ship Dauntless with a weapon that made the crew feel as though they "... *were being compressed, not as a whole, but atom by atom...twisted...extruded...in an unknowable and non-existent direction*". They were no longer in the space that they knew and speculated that they "...*wouldn't have surprised me if we'd been clear out of the known universe. Hyperspace is funny that way...*" . In addition, a weapon known as a 'hyperspatial tube' is mentioned and used by the Boskonians and their allies the Delan's to attack earth. It is described as an 'extradimensional' vortex. the terminus of such a tube cannot be established too close to a star due to its apparent sensitivity to gravitational fields. In 1947, *Children of the Lens* by E.E. 'Doc' Smith describes the planet Boskonia attacking Telus by sending a fleet of warships through a 'hyperspacial tube' instead of through normal space. This transport method is regularly used by Palainians and was not invented or discovered by earthlings.

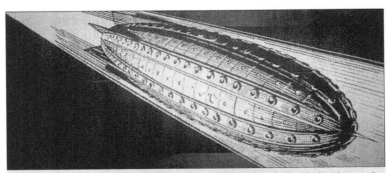

The "Dauntless" flying down the hyper-spatial tube. (Astounding, Jan 1940. Artist: Charles Schneeman.)

Hyperspace

In 1940, Nelson Bond had also alluded to hyper-space in a story *The Scientific Pioneer Returns*. Using a 'velocity intensifier' powered by hypatomic motors, a ship accelerates into 'imaginary space', which turns out to be a parallel universe. Exceeding the speed of light in normal space is impossible, and instead of traveling to a distant point in normal space, the ship is thrown into another universe entirely. *"...Einstein and Planck fiddled around with hyper-spatial mechanics and discovered that mass is altered when it travels at high velocity. The gadget worked better than you expected."*

An attempt was made several times in the 1950's to combine the limitations of relativistic travel in 'normal' space with the perceived unlimited possibilities of inter dimensional travel. In *Citizen of the Galaxy* (1957) by Heinlein, and *The Stars Like Dust* by Isaac Asimov and published in 1950, for example, a space ship would first have to boost to near-light speed in normal space using atomic motors, before it could make the transition into hyperspace. FTL travel is acknowledged to be impossible in normal space. Great expenditures of energy are needed to enter this 'space within space'. Traveling into and out of hyperspace can be a tricky, and even deadly, process.

Overall, the secret to interdimensional travel is an extremely difficult one to discover. It represents a technology and understanding of the physical world that in many instances only mysterious super-civilizations possess. Earthlings are reduced to passively using this technology without really understanding it. Journeys through hyperspace are not willy-nilly. The trajectories taken must be computed beforehand with considerable care. Entry into hyperspace can occur by a variety of means: The sudden unleashing of natural forces; the application of powerful

Interstellar Travel

magnetic fields; traveling to places in space where natural 'incongruities' exist, or the application of the emanations of mysterious new elements.

In *Starman Jones* (1953) by Robert Heinlein describes how the transition into something called 'N-space' was a delicate matter requiring careful calculations. At some points in interstellar space, space was folded over on itself in 'Horst Anomalies'. These gentle foldings of space do not represent 'warpage' so, apparently, do not cause unusual gravitational fields in empty space. This makes them difficult to locate. We learn that to attempt to travel faster than light speed causes a ship to 'burst out' of normal space. If the ship does this carefully near a Horst Anomaly, however, it is merely shifted to a distant point in normal space. Anomalies have to be carefully mapped by exploration ships to find out just where the journey ends. Astrogation consists of putting the ship at exactly the right spot in an Anomaly, with exactly the right velocity and acceleration to insure that the Jump ends up where you want to go. The relationship between 'N-space' and hyperspace is a bit unclear, but they function in similar ways in the story so we can probably assume that they are equivalent.

Some other perils of hyperspace are pointed out by Milton Lesser in his 1950 short story *All Heros are Hated*. The year is 2900 AD, interstellar travel has been a commonplace for centuries. Travel to the Magellanic Clouds takes a few years, and a hop to the colonies around Fomalhaut takes 6 days; at least until the 12 billion inhabitants of these 6 worlds were annihilated. The spaceship Deneb exited hyperspace with its drive still turned on by the time it entered normal space. This

Hyperspace

caused Fomalhaut to go nova and incinerate all life in this planetary system! An even more terrifying possibility is described in Alfred Bester's *The Push of a Finger* (1942). By creating an 'osmotic spatial membrane' scientists are able to tap a limitless source of energy from hyperspace. This energy, however, begins to drain into our universe causing our universe to come to a premature end. Fortunately, this event was stopped by a time traveler from the future who interceded at just the right moment!

In the *Foundation* series by Asimov, hyperspace travel in 'Gravitic ships' had to be made far from strong gravitational fields otherwise the calculations became progressively more difficult and physically uncomfortable for the human cargo. These ships isolate themselves completely from external gravitational fields. Asimov's epoch of the fall of the galactic empire also presents us with the scenario that the secret of FTL travel, and the building and servicing of such ships, is a skill that can be easily lost to a civilization.

Between 1928 and 1940, the magic of hyperspace quickly became the favorite mode for FTL travel that circumvented completely the ordinary relativistic prohibition against FTL travel in 'ordinary' space. From this, vast galactic empires and sprawling epochs of adventure were created almost overnight. With few exceptions, the need for explaining the details of hyperdrive became less intense as the story lines were developed with ever increasing depth and complexity. With the entire galaxy as a stage, the scale of human science fiction imagination grew by orders of magnitudes.

Interstellar Travel

In the imagination of Larry Niven as well, humans did not invent FTL travel as discussed in *The Borderland of Sol* (1974), but had to buy this secret, along with entire preassembled spaceships from a civilization known only as the Puppet Masters. Evidently, there are many levels of hyperdrive. The unlucky space traveler that enters a strong gravitational field can easily get scattered into one of the other hyperdrive levels and never find their way back to normal space.

Jerry Pournell also developed a new mode of transportation in *He Fell into a Dark Hole* (1974). A new 5th force discovered by the physicist Alderson is found to have a 5th dimensional component to it, and is produced in all nuclear reactions. Every star becomes the node for a complex network of field lines called 'tramlines' in the Pournell-Niven novel *A Mote in God's Eye*, which voyagers may take advantage of since the Alderson

Hyperspace

force propels the traveler along the tramlines through hyperspace. Again, expedition ships have to identify the destinations for each tramline emerging from the sun. When the ship arrives near its intended destination, the local gravity snaps the ship out of hyperspace and back into normal space. Arthur J. Burk's *The First Shall be Last* also describes spaceships that travel along 'gravitic lines of force' from planet to planet at nearly light speed. The solar system is a complex webwork of lines connecting each planet and crisscrossing interplanetary space. Navigation is a complex matter of starting out on one line and switching over to others to get to the desired destination. These lines do not run straight. Similar accounts of mysterious currents flowing through space are found in Raymond F. Jone's *Correspondence Course*(1945). These 'magnetic currents' can be ridden by space ships to various destinations.

Arthur C. Clark's 1968 novel *2001: A Space Odyssey* is a less complicated introduction to hyperspace travel that involves mysterious black 'monoliths' built by a super-civilization, which act as gateways to other locations in the galaxy or the universe. It is never actually made clear whether the journey by Bowman is to another location within our spacetime, or if he actually leaves our universe entirely. No details are given how the transfer occurs from place to place, or what role the monoliths serve in this process. The novel by Carl Sagan *Contact* (1985) is a story that runs along similar lines to *2001: A Space Odyssey* in terms of FTL technology. A message we receive from a super-civilization gives us the instructions for building a vessel, creating a 'dimple in spacetime' to which alien engineers may attach a 'wormhole' bridge. The voyagers find themselves

Interstellar Travel

traveling across space to the center of the galaxy after a stop-over at Vega.

A similar story is *Age of Miracles* by John Brunner published in 1965. Earth is invaded by dozens of 'cities of light'. An advanced civilization has decided to set up a local node for their interstellar transportation system in our solar system. The 'cities' are not made of matter as we know it, but 'slowed down coagulations of energy'. Their interiors are twisted into higher dimensions and result in disturbing sensory shifts to any unshielded human who enters. Eventually, earthlings find themselves free to use these doorways to travel much as rats cross the Atlantic Ocean on ships. The aliens are indifferent to our invasion of their 'subway' system.

More aggressive journeys into alternate Realities are explored by Michael Moorcock's novel *The Sundered Worlds* (1966) which is probably one of the most detailed excursions of its kind. During the 26th century, human civilization extends to the limits of the galaxy thanks to hyperspace drive. To navigate through hyperspace and the other 'alien dimensions to space time', rare individuals called 'Space Sensors' with ESP-like talents are used. Renark was one of these, and the hero of the story. He meets up with strange humanoid beings from the galaxy Messier 31 who tell him that the universe will re-collapse in a few years. To save humanity, Renark must find a way for humans to leave our 4 dimensional continuum completely. He heads for a strange solar system called the 'Shifter' whose orbit is at right angles to the rest of spacetime and which passes through our universe every few hundred years. Our 4-dimensional universe coexists with an infinitude of other universes in multi-dimensional space; a view

Hyperspace

which is called the 'Multiverse' theory. Like the separate pages in a book, each continuum has its own laws and indigenous intelligent race. In the Shifter, Renark meets the beings called the Originators. The Originators are multi-dimensional beings who created and maintained the multiverse as a nursery for a lifeform to replace them and keep Reality from decaying into chaos. Renark finally learns from them the secret of the 'inter-continuua drive' and saves humanity. The Originators then evolve humanity to serve as their replacements, and thereby save the Multiverse from destruction.

In *The End of Eternity* (1955) by Asimov an even more complex tale is woven in which in the distant future, humans have learned how to exist outside normal spacetime. They then set about making minor alterations to the unfolding of human history in order to minimize strife and maximize human progress. This story spans millions of centuries and develops through the guidance of computer calculations of probabilities for various Earth histories. Technicians re-enter normal spacetime and make subtle adjustments whose propagating effects multiply down earth's timeline and lead to the correct, desired result. A barrier in spacetime is discovered millions of years in the future, which is caused by even more advanced humans who are trying to protect themselves from the worldline tampering going on during these earlier ages. This future Earth civilization and discovered that they lived on a very low probability worldline for human history and used this fact to uncover the tamperings during the early ages. A similar story is Fritz Leiber Jr's *Destiny Times Three* (1945). Eight humans obtain possession of the 'Probability Engine' which is a super-mechanism operating outside spacetime. In secrecy, they use it

Interstellar Travel

to split time and create alternate histories for the earth, allowing only the best to survive.

Another story about alternate time streams in found in A.E. van Vogt's *Recruiting Station* in 1942. Future earthlings from 200,000 years from now recruit 20th century earthlings to fight a war against a second group of earthlings from the rest of the solar system. The terrestrial time stream is manipulated to create 18 alternate solar systems in which the battles over political control are waged. Eventually, a more advanced race of humans from the 4900th century step in to save the embattled underdogs against domination by the hostile pseudo-fascist party.

Patterned after Arthur C. Clarkes' *Rendezvous with Rama*, Gregory Bear's *Eon* and *Eternity* novels begin with the entry into our solar system of a hollowed-out asteroid. The drama unfolds rapidly when the human explorer's discover that although its external length is only a few hundred kilometers, inside it extends billions of miles! Built by earthlings called Geshels, 13 centuries in our future, it is fashioned out of artificially twisted spacetime. This cylindrical world called the 'Way' is an entry corridor into superspace. Every thousand kilometers of travel down its axis represents a time shift of one year in history. Within each fraction of a millimeter along the axis are 'stacked' individual spacetime geometries representing the alternate possibilities. Artificial gateways can be created into these alternate spacetimes and searched for habitable versions of the earth. The walls of the way are described as being formed from 'jammed-up probabilities' from alternate universes attempting to take-on a particular state, or in the slang of the human geometrodynamicists 'superspace trickery'. Its creation, however,

Hyperspace

turns out to be a disaster to the harmonious laws of superspace and, ultimately, must be destroyed.

It took several decades for this to become an established principle of SF fact, but it is now widely recognized that so long as we travel through ordinary space, we must abide by the limitation that we cannot travel faster than light speed. Few of the spaceships carry their own reaction mass with them to reach these high velocities, which means that in the world of SF, humans have learned how to finesse the laws of physics to avoid the conservation of momentum difficulty. Arthur C Clark's 'Quantum Drive' or Poul Anderson's 'Interstellar ram jet' are plausible basis for such ideas.

If the technologies in the worlds of SF can overcome the need for reaction mass, then the door is wide open for what you might call 'wonder drive' technology, which has no basis in the workings of the known universe, although they sound plausible.

The underlying common thread behind nearly all FTL or trans-dimensional technology is that space drives can produce the appropriate conditions needed for them to operate including 'space warps', 'new forces' or 'phase shifting'. They do so by expending only a miniscule portion of their own rest mass. Ordinarily, enormous energies have to be marshaled by the universe to create localized space warps called black holes whose sizes are only a few kilometers. If we were to use black holes as a probe of the relevant physics, to warp space at a scale of 100 meters in order to accommodate a small spaceship, this would require expending 3% of the sun's rest mass! The resulting gravitational field would shred all forms of matter into their

Interstellar Travel

constituent elementary particles, and although a 'throat' might be created, so too will an event horizon. Only a 'Quantum Drive' has the potential of harnessing this much energy since the quantum fluctuations of space that provide the energy contribute 10 to the 60th power solar masses per cubic centimeter of space. The manner in which such energy would be converted into a useable 'hyperspace doorway' is unknown to SF authors.

Clearly, in the world of SF we are currently living in the brute force era of space technology where we do not as yet know how to manipulate space and its structure without expending vast amounts of energy and at the same time, having to deal with lethal gravitational fields.

It is also taken for granted that the passage out of our continuum and into hyperspace will be comfortable enough for humans to travel through. By some means, future science will have mastered the ability to create large 2 meter to 100 meter class wormholes while holding at bay the enormous gravitational fields that such openings in spacetime invariably represent. Without proper shielding, humans and ships alike will be crushed or fractionated. Only for some authors is this travel described as unpleasant or psychologically upsetting, and no special precautions are needed to survive the gravitational forces.

SF written during much of the Space Age appears to be comfortable with the notion that spacetime is a 'fabric' that we will eventually learn how to work as in *Eon* or *Contact*. Travel through space will still be limited to sub-light speeds, but the conviction seems to be great that a shortcut around this

Hyperspace

annoying limitation will be found. With modern technology and scientific understanding, it is infinitely easier to part the waters of the ocean and to walk upon air, than to alter the geometry of space. But perhaps, one day, just as we readily create magnetic fields from electrical fields, we may discover how to convert electric fields into gravitational ones, without at the same time vaporizing our laboratory equipment.

Science fiction has been with us as a recognizable literary genera for nearly a century and represents an evolving network of ideas that develop almost parallel to revolutions in scientific thinking. Jules Vern's submarines and airships were almost patentable. Then came atomic-powered rockets of the Buck Rogers variety, followed by a progressive refinement of drive technology into warp engines, hyperdrive and teleportation. As the technology of SF has become more sophisticated, it has also found itself more in the league of magic. It has all but left the real world, or reasonable extensions of it. Only the setting (the Galaxy) and the human condition (greed, power, love, war) remain as fixed reference points operating in recognizable ways. Has SF finally evolved beyond its own definition? Unlike the science fiction of the past century, modern SF provides no satisfying linkage between what we know today and a plausible route to the technology of the future. The old Maine saying *'You can't get there from here'* applies to nearly all of the SF worlds the genera is currently obsessed with. Without this linkage, SF has perhaps unintentionally transformed or evolved itself into the category of Fantasy; a landscape also populated by magical solutions to physical problems. Arthur C. Clarke's Third Law injunction that a sufficiently advanced civilization will have technologies that are imperceptible from magic might ameliorate this difficulty.

Interstellar Travel

However, it is probably just as well that SF as such masters as Arthur C. Clarke to serve as a necessary regulator on unbridled scientific fantasy!

To escape the latest run-in with hostile aliens, Captain Kirk of the starship Enterprise orders Engineer Scotty to take the ship to 'warp factor 6'. The mighty engines open a doorway into the magical world of hyperspace and in an instant, the Enterprise is taking a short cut through space. A trip to the nearest star now takes only a minute or so, not centuries, and Captain Kirk survives to continue his five-year mission. Yet there is not one of us who has never been inspired by the awesome possibilities opened up by FTL travel, the opening up of the galaxy to human commerce and colonization. Whether we will ultimately be able to create furniture from curved space, partake of a multidimensional reality, or directly view all of humanities alternate histories, becomes less of an issue than being able to fuel the imagination with these endless possibilities.

Science fantasy has always dealt with fantastic ideas at the very limits of believability, but sometimes the distinction between science fact and science fiction can become murky indeed. "Space...the final frontier" is a truism that takes-on a very different meaning in light of what is now known or suspected about space. Today, physicists and astronomers are exploring exciting new ideas that may be the basis for a future at least as exciting as Captain Kirk's 23rd century.

Hyperspace

But we should also consider that the technology needed for interstellar travel may be so difficult, and so costly to create that we may never achieve it ourselves. Instead, as in many science fiction stories such as *Contact*, *Babylon 5*, *Stargate* and *Interstellar*, some benevolent aliens may have to give us the technology instead!

Interstellar Travel

The True Shape of Space

Of course 'outer space' is filled with planets, stars and galaxies. Without these ingredients space would be a sterile nothingness. But a simple glance at the infinite blackness of the night sky shows that this intangible ingredient is one of the most common in the physical world; It is also the least understood. Unlike most things that can be directly experienced, it is difficult to speak critically about something you can neither touch nor see. Fortunately, when it comes to pure, empty space, the situation is not quite so bleak.

As we all learned in high school, one of the most basic properties of space is its extension, also called its dimensionality. The world in which we work, pay taxes and take vacations is a three-dimensional one. In it, we are free to move forward and backward, side-to-side, and to a limited extent, up and down. The intricate gyrations of a ballerina or a gymnast tell us that no matter how we might move, we never 'turn a fourth corner' and find ourselves moving along a new direction through space. This tells you right away that Captain Kirk's world of hyperspace is not going to be an easy one to break into, especially by something as big as a human or a starship!

And then there's Time.

A strictly three-dimensional world is pretty boring. Nothing happens in it. Suppose you tell your friend that you will meet her at the entrance to the Washington Monument. Your well intentioned instructions will help her narrow your location in the universe to a six foot cube of space at a particular point on

Interstellar Travel

the surface of Earth. But unless you also say when to be there, the instructions are useless. Time is a vital fourth coordinate or dimension to our world. Without it, we would all be trapped in a perpetual Now, much like the frozen images captured on a photograph. Space and Time taken together define the complete arena in which we live. They form such an integral, cohesive framework for our existence that physicists since Albert Einstein refer to their combination as simply 'spacetime'.

Spacetime is vast. It extends well beyond the Earth and solar system, encompassing the entire universe out to the farthest galaxy. Its indivisible time-like aspect also extends from the instant that the universe flashed into existence, through the present moment, and on into the future.

Where did spacetime come from? Astronomers who study the universe have developed a detailed model of its evolution called the Big Bang Theory. About 14 billion years ago, everything in the universe came into existence in an awesome explosion. The feeble light from the fireball of creation can still be detected by sensitive instruments as they peer into the depths of space. The magnitude of this event is truly mind boggling. Earthbound explosions begin with a bomb whose detonation sends debris flying out into space. But in the Big Bang, not only did matter come into existence, but space and time as well!

According to some recent theories, before the Big Bang, our particular spacetime simply did not exist. Anywhere. Anywhen. It is difficult and somewhat troubling to imagine that time and space had a beginning. Even among religious cosmologies, both ancient and modern, this has been a common theme.

The true shape of space

More amazing and profound than its scope and origin are new discoveries that may portend even more remarkable revelations about the nature of spacetime. These discoveries have come not from the study of the grand design of the universe, but from a

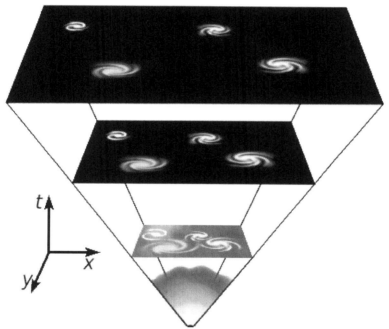

meticulous investigation of the composition of matter spanning over 300 years of experimental and theoretical work. From this intense human activity has emerged a detailed understanding of just how matter and force come together in spacetime to build-up the complex structures in our world.

At some time in our schooling we are told that matter consists of atoms; one for each element like oxygen or iron. The atoms

Interstellar Travel

themselves are built from even more elementary particles called electrons, protons and neutrons of which the latter reside in the dense atomic nucleus. Since the 1960's, gigantic machines commonly called 'atom smashers' have uncovered an even finer structure to matter. Neutrons and protons are, themselves, made from minute particles called quarks. All common forms of matter can now be represented by the combination of just three particles: one electron and two kinds of quarks.

But there is more to Nature than matter. Without forces such as gravity, the world would be formless and devoid of living matter. Once again, although the dynamics of the world seem bewilderingly complex, there are only four distinct types of forces in Nature, each playing its own crucial role in orchestrating the universe.

The force of gravity acting over billions of years assembles matter into galaxies and stars, choreographs the dance of the planets around the sun, and keeps our feet planted firmly on the ground. The electro- magnetic force holds electrons captive inside atoms and allows matter to give off light for us to see. The strong nuclear force binds atomic nuclei together, and its release in nuclear fusion keeps the sun and stars shining. Last but not least is the weak nuclear force which causes matter to decay, and stars to detonate as supernovae in devastating explosions. Detailed mathematical descriptions are available for each of these phenomena which allow anyone interested in such matters to comprehend and perceive the physical world with unprecedented clarity. Basic phenomena in the world, from the color of a sunset to the birth a star, are no longer regarded as

The true shape of space

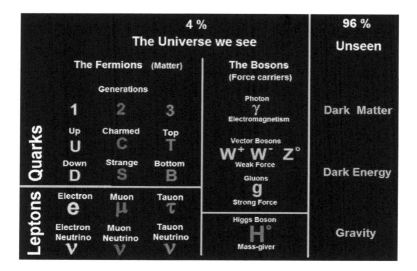

capricious and mysterious, but can actually be predicted with fair accuracy.

Physicists, however, want to do more than merely describe how each separate force acts upon matter. Physics is more than merely the passive 'high-tech' bookkeeping of Nature's comings and goings. It is a search, guided by experiment, for the basic, universal principles that underlie how the physical world operates at every imaginable scale, from the most distant galaxy to the innermost workings of the atom. In creating such a comprehensive 'Theory of Everything', somewhere along the way one of the greatest remaining challenges to our understanding of the physical world must be faced. A glimpse of this challenge can be seen by thinking about a simple electron. In this image, created by a scanning tunneling microscope, we

Interstellar Travel

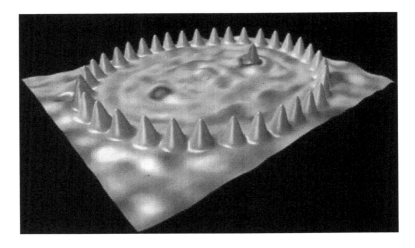

see 39 cobalt atoms assembled in an elliptical ring with two foci. At atom (top) placed at one focus creates a mirror of itself at the second focus due to the wave-like properties of quantum matter.

If you were to draw an imaginary line through space, piercing the center of an electron, why is it that you single out one of these points as an electron but call all of its neighbors 'empty' space? It is easy to semantically define them as being different, "This one is the electron, that one over there is space", but how do you go about handling this difference quantitatively?

Many schemes for describing the essential difference between matter and empty space have been tried over the decades; many have failed. The electron was at first thought to be a tiny sphere of matter whirling around the nucleus of an atom like a miniature planet. As intuitively seductive as it was, this idea fell into disrepute once Albert Einstein developed the Theory of Relativity. Then the revolution of Quantum Mechanics showed

The true shape of space

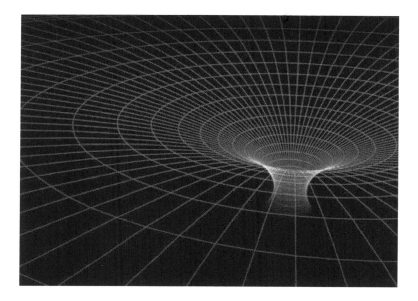

that all matter had wave-like properties; electrons at a particular instant were not located at fixed positions in space, but seemed to be in many places at once.

For the last 50 years, electrons and other elementary particles like quarks are routinely thought of as small dots of pure energy whose boundaries vanish into the undefined fabric of space itself. It isn't that physicists have directly measured this to be the case, only that this is the only remaining working model for the electron that has survived, and is consistent with all that we know about electrons, both theoretically and experimentally.

Theoreticians since Einstein have speculated about the geometric features of spacetime, and the structure of electrons and matter for decades. The growing opinion now seems to be

Interstellar Travel

that, ultimately, only the properties of space such as its geometry or dimensionality can play a fundamental role in the defining what matter really is. In a word, matter may be just another form of space. If the essence of matter is to be found in the geometric properties of 'empty' space, our current understanding of space will not be sufficient to describe all of matter's possible aspects.

Remember, you needed to specify four coordinates in order to meet your friend at the Washington Monument at the right time and place. For most things in the world, including the motions of the planets, stars and galaxies, four dimensions is enough. But to describe the world of elementary particles, physicists have to add some additional coordinates to spacetime to keep track of the properties of subatomic particles. There does not seem to be enough room in a strictly four-dimensional universe to explain why matter looks and acts the way it does. Some of the most promising theories require that the universe exists in as many as 7, 10 or even 26 dimensions at once!

These added dimensions do much more than just tell where a particle is located in the universe. They actually determine how that particle will look! According to Superstring Theory which was developed by physicists John Schwartz and Michael Green in 1982, every point in spacetime is represented by its usual four coordinates, along with up to 22 more 'stunted' coordinates. Every particle, on the other hand, is given an address in this 26-dimensional universe; an address telling whether the particle is located in a star or in the paper you are reading. The additional coordinates tell us what kind of a particle it is. Depending on how a particle 'moves' in these other directions it might, for example, be seen as an electron, a quark, or even 'empty' space.

The true shape of space

How can a particle's motion along a particular dimension change its character so drastically? Fortunately, you don't have to be an expert in physics to get some idea of how this might happen. For instance, your entire life's history, stretching along the fourth dimension of time, contains versions of you that are an infant, a young adult, or a senior citizen. Now suppose that we could move freely through time, we would be able to witness your drastic physical transformation from one kind of human 'particle' to another! By knowing your position along the fourth dimension the rest of us can keep track of how to interact with you. At some point it will, of course, be better to interact with you using 'baby language' than at other times!

We know from the gyrations of ballerinas and gymnasts that, if they exist at all, these dimensions can't be very big. There is no danger of taking a walk to the store and suddenly finding yourself in the 17th dimension!

Interstellar Travel

Only subatomic particles like electrons are small enough to gain any benefit from such a journey. Superstring Theory and some of its predecessors say that these added dimensions are rolled up into miniscule balls one trillion trillion times smaller than an atom. An atom would have to be enlarged to the size of the Milky Way galaxy 100,000 light years across, before any signs of them would be apparent!

Just as the lumps and bumps in the geometric shape of a piece of paper will control the motion of marbles moving across them, there is also an intimate relationship between the geometry of a 26-dimensional spacetime and the behavior of matter. Since the properties of the elementary particles and forces in the universe are already known, this can be used to discover what kind of geometry a 26-dimensional universe would have to have in order to resemble the universe we are familiar with.

The answer to this question is still being searched for today. But if and when it is found, it is believed that the geometric basis for the Theory of Everything will at last have been uncovered. To say that this will be a major accomplishment is an understatement. In fact, some physicists like Stephen Hawking in *'A Brief History of Time'* even predict that the discovery of the correct geometry will herald the end of physics as we know it.

It has been said that to understand the motions of the clouds in the sky, you must first study the winds and currents of the invisible atmosphere itself. Like clouds, it may well be that matter is merely a tracer of activity at a more basic level in the physical world. The deep roots that elementary particles have

The true shape of space

may reach down into the bedrock of spacetime whose geometry ultimately controls their properties and how they are destined to interact with one another. Like an oak or a maple tree, we measure and perceive only their broad canopies. Their roots remain forever hidden.

It would seem that if our modern theories are correct, Captain Kirk will have to do something else other than duck into hyperspace to escape his enemies! Because the universe may manifest these other 'hyper' dimensions at the subatomic scale, only electrons can take advantage of them. Few of us would especially enjoy being squeezed to the size of an electron to escape even the most hostile alien! Although such rapid travel through hyperspace may never be possible given the rules upon which our universe may be based, there are other even more exciting possibilities.

If matter is 'simply' twisted space, would it be possible to create matter and perhaps even entire, artificial mini-universes out of warped space? Though technically difficult, some physicists such as MIT's Alan Guth have seriously thought about these possibilities and consider them within the realm of possibility in the distant future. Although it may not be possible for humans to travel through any of the other dimensions to space as a short cut to some distant star, what about the massless particles of light called photons? Could we at least send radio messages through hyperspace almost instantaneously, and not have to wait hundreds of years to get a reply from our colonists orbiting Antares? Then again, just because the physical world may not naturally include something like hyperspace on the interstellar scale, is it possible that no injunction exists forbidding its

Interstellar Travel

artificial creation? Perhaps given enough raw energy focused on a small enough region of spacetime, many natural barriers could be overcome.

The scientific exploration of the world has taught us much about the way the universe is put together. We are now familiar with nearly all of its most important rules and regulations for living in harmony with its basic phenomena. It is always difficult to predict where the next great revolution in thinking will come from. Perhaps some of the current ideas about matter and space will not even survive the end of the 20th century as new experiments are developed. Then again, the scientific advancement of the last three centuries would have been impossible had not some ideas, as farfetched as they seemed at the time, been correct in one form or another!

So, the next time you gaze at the night sky, take a moment to reflect on the nature of the vast emptiness between the stars. Even perfectly empty space, the quintessential nothingness, may be a far more sublime and complex ingredient to our universe that we have ever before imagined!

Where would we go?

Before 1992, there was no evidence that planets existed anywhere except within our own solar system. There were countless books and research papers written about the probability that planets existed around other stars, and there were even a few efforts made to try to detect them by looking at the wobble of a star as it traveled through space. One of the most exciting false-alarms happened during the early years of the Space Age. In 1965, an exciting announcement from astronomer Peter van de Kamp claimed that Barnard's Star only 6 light years from Earth had an orbiting planet. Unfortunately, this was followed in the 1970's by improved studies that failed to show this purported planet. By the 1990s, observations with the Hubble Space Telescope eliminated planetary candidates around Barnard's Star that were more massive than our planet Jupiter orbiting with periods of less than 1000 days. In addition, the data also excluded any planets inside the stars habitable zone with masses more than 8 times that of Earth. These limits, however, still do not completely exclude an Earth-like planet orbiting this star closer than the orbit of Jupiter in our own solar system.

First, a bit about units of distance measure. Astronomers famously use the light-year which is the distance light travels in one year. Since the speed is 300,000 km/sec and there are about 31 million seconds in one year (3600 x 24 x 365.25) you get about 9.5 trillion kilometers. Another unit often used is the parsed, which is 3.26 light years or just 30.8 trillion kilometers. Finally for solar system and interplanetary distances we use the Astronomical Unit (AU), which is the distance between Earth

Interstellar Travel

and the Sun. The adopted value is about 149.6 million kilometers. By the way, at a distance of one parsec, the angle subtended by 1 AU is one second of arc, which is of course 1/3600 of a degree of angle. The distances to the planets in our solar system are: Mercury (0.39AU), Venus (0.72 AU), Earth (1.00), Mars (1.5 AU), Jupiter (5.2 AU), Saturn (9.5 AU), Uranus (19.2 AU) and Neptune (30.1 AU). Also, the orbit time for Mercury is about 88 days, while Neptune is about 165 Earth years. This gives you a sense of how big our solar system is compared to the planets we will now have a look at!

By the early 1990s, the idea that planets orbiting other stars still lacked any supporting evidence other than a general feeling that planets probably do exist but they are just too hard to detect directly. But then in 1992 Aleksander Wolszczan and Dale Frail announced the discovery of a multi-planet planetary system around the millisecond pulsar PSR 1257+12. These were the first two extrasolar planets confirmed to be discovered, and thus the first multi-planet extrasolar planetary system discovered, and the first pulsar planets discovered. Two additional planets of lower mass were later discovered by the same technique. The closest planet had a mass of only 2% of our Earth, and orbited just under 0.2 AU from the pulsar with a period of 25 days. The next planet was about 4.3 times the mass of Earth and orbited at 0.36 AU with a period of 66 days. The outermost planet had a mass of 3.9 times our Earth at a distance of 0.46 AU and an orbit period of 98 days. The entire planetary system could roughly fit inside the orbit of Mercury, but would be bathed not by the warmth of a normal star, but the harsh and even deadly radiations of a spinning neutron star.

Where would we go?

If the planets had existed before the star went supernova, they would have been completely sterilized. But there is another problem too. This planetary system is located about 1000 light years from Earth towards the constellation Virgo. If this were the only example of a planetary system in our corner of the Milky Way, to get there and explore its planets would be a monumental challenge for any means of interstellar travel. Luckily this is not the case!

On October 6, 1995 astronomers Michel Mayor and Didier Queloz from the Observatory of Haute-Provence in France announced in the journal Nature the discovery of a planet orbiting the star 51 Pegasi. This star is at least 6 billion years old and is a G2.5 star, making it very sun-like in temperature, color and mass. The planet itself is no prize. Nicknamed Bellerophon, the planet is at least half as massive as Jupiter, orbits its star once every 4 days and has an estimated surface temperature of 1800 F. In fact, it seems to be losing its atmosphere and literally evaporating before our eyes! But the good news is that it is located about 50.9 light years from Earth, which means that it is close enough to visit with some effort. But even better than this, this discovery validated the idea that we now had technology good enough to actually detect planets orbiting stars! Between 1995 and 2009 the race was on to detect more 'exoplanets' using even more sophisticated, ground-based equipment.

Then on March 6, 2009 NASA launched the Kepler mission, which was specifically designed to search for transiting planets among 150,000 stars it would monitor every 42 minutes around the clock. This is similar to the Transit of Venus observed by the Solar Dynamics Observatory in 2012.

Interstellar Travel

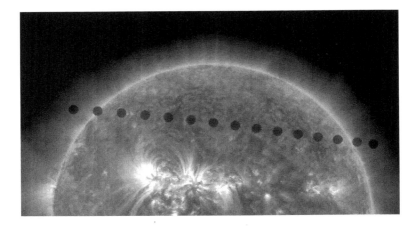

On January 4, 2010 the mission announced the detection of its first five exoplanets, all much larger than Jupiter and orbiting so close to their stars that they resembled Bellerophon. The stars were also located over 1000 light years from the Sun, making this first collection of planets a very hard journey to visit.

Prior to the Kepler mission, most confirmed planets were gas giants comparable in size to Jupiter or larger as they are more easily detected. But the hundreds of detected 'Kepler exoplanets' are mostly between the size of Neptune and Earth. As of March 2015, a total of 1901 confirmed exoplanets are listed in the *Extrasolar Planets Encyclopaedia*. That count includes 1199 planetary systems, of which 478 are multiple planetary systems.

The table below shows how these confirmed exoplanets were grouped according to their masses in September 2014. These are candidate objects of which 1000 have been confirmed by independent studies of their transits through other means such as ground-based telescopic monitoring.

Where would we go?

Class Name	Radius Range (Re)	All Candidates	Insolation 0.32-1.78
Terran	≤ 1.25	1025	14
Superterrans	1.25 < R ≤ 2	1312	44
Neptunians	2 < R ≤ 6	1577	109
Jovians	6 < R ≤ 15	274	25
Superjovians	R > 15	374	24
	Totals:	**4562**	**216**

Exoplanets confirmed by other means tend to be found closer to Earth and are included in the *Extrasolar Planetary Encyclopedia*, which lists 1909 planets detected and confirmed by April 2015. The nearest known Earth-sized planets from these surveys are listed in the following table:

Planet	Discovery	Distance	Mass	Orbit
Alpha Cen Bb	2012	4.3 light yr	1.0 Earth	3.2 days
GJ 876 d	2005	15.3	5.6	1.9
HD 20794 b	2011	19.6	2.6	18.3
HD 20794 c	2011	19.6	2.3	40
HD 20794 d	2011	19.6	4.7	90
GJ 581 e	2009	20.4	1.9	3.1
HD 581 c	2007	20.4	5.3	12.9
HD 581 e	2009	20.4	1.9	3.1
GJ 667 Cb	2011	23.6	5.6	7.2
GJ 667 Cc	2011	23.6	4.0	28
61 Vir b	2010	27.9	5.0	4.2
GJ 433 b	2013	28.9	5.6	7.4
HD 85512 b	2011	36.2	3.4	58.4
55 Cnc e	2004	40.1	8.1	0.7
HD 40307 b	2009	42.4	4.0	4.3
HD 156668 b	2011	79.9	4.0	4.6

Interstellar Travel

What we see is not that the number of Earth-sized planets decreases with distance from Earth, but that the method for finding these small planets begins to break down as we study stars farther from Earth. Also, the easiest planets to detect are very close to their stars where their gravitational forces are strongest. Detecting Earth-sized and smaller planets between the orbit of Mercury and Mars in these systems (orbit periods between 50 days and 800 days) is very difficult and has to await future technologies to detect them, if the planets exist at all.

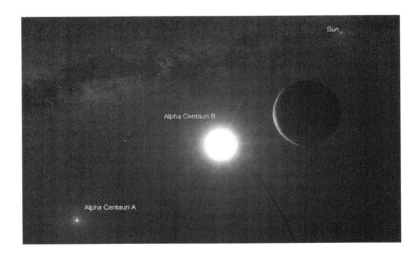

Habitable Zones

As important as the size of the exoplanet might be in determining its gravity, we also want to make sure that the planet has a fighting chance of having liquid water on its surface. If it's an ice-bound world or a Venus-like inferno, what good is it to interstellar travelers? This leads us to the second important astronomical issue for exoplanet destinations: Is the planet in the habitable zone of its star? In other words, is the surface temperature of the planet somewhere between the freezing point of water (32 Fahrenheit, 0 Celsius or 273 kelvin), or the boiling point of water (212 Fahrenheit, 100 Celsius or 373 kelvin)?

The two most important things that control a planet's temperature are its distance from its star, and the luminosity of its star. Luminosity is just the number of watts of radiant light energy that the surface of the star emits. Our sun, for example, produces 3.8×10^{26} watts. Dimmer stars can produce 1/10000 of this amount, while supergiant stars can produce a million times the solar value. Distance and luminosity then determine how many watts per square meter arrive at the planet's surface. For our Earth, it's easy to calculate this number. In January this 'Solar Constant' is 1400 watts/m² when Earth is closest to the sun, and in July it is 1300 watts/m² when Earth is farthest from the sun. This is how much light

> Imagine a sphere whose surface is at the distance of Earth's orbit. From $S = 4\pi D^2$, and $D = 150$ billion meters, the area of this surface is 2.8×10^{23} square meters. Dividing the solar luminosity in watts by the area you get 1344 watts per square meter.

Interstellar Travel

energy arrives at the top of Earth's atmosphere. At the surface of Earth, this is called the solar insolation, which is responsible for heating the surface and driving plant photosynthesis.

The next most important thing is, how much of the solar constant does the planet's surface actually absorb? Astronomers call this the planetary albedo, represented by the letter A. A maximum value of A=1.0 means the surface acts like a perfectly-reflecting mirror. A minimum value of A=0.0 means the surface acts like the blackest-possible object and absorbs 100% of the in-falling energy. Fresh snow is about A=0.9 on this scale while our Earth, with its continents, oceans and icecaps has an average albedo of about A=0.3. Worn street asphalt has an albedo of A=0.1. This is why snow feels so cold, because it is emitting just about everything it receives and stays cool, while asphalt feels very hot because it is busily absorbing nearly all the radiation that touches its surface!

After this, we have to consider whether the planet is rotating or not. If it is not rotating, the solar radiation that reaches its daylight surface covers the area πR^2, and when the radiation is emitted back into space, it is also emitted over this same area. But when the planet is rotating, the radiation that comes in is emitted over the entire surface area of the planet which is $4\pi R^2$. So for a rotating planet like Earth, the actual surface heating and cooling radiation is just ¼ the solar constant or about 340 watts/m². This is the number that drives Earth's climate. Without an atmosphere, Earth would still get 340 watts/m², but it would reflect back into space A=0.3 and absorb (1-A) or 0.7 of this energy.

Habitable Zones

The final piece of information we need is that a surface heated to a temperature of T in kelvin units will emit a precise amount of energy because of the Stefan-Boltzman Law, which is written in the compact formula $F = 5.7 \times 10^{-8} \, T^4$ watts/meter^2.

When we set F equal to the solar energy absorbed by the surface of Earth at its distance from the sun, and with its albedo, we can solve the equation to get the distance D that a planet will be from its star to have an average surface temperature of T in kelvin units. By setting T = 373 K

$$5.7 \times 10^{-8} \, T^4 = \frac{(1-A) \, L}{16 \pi D^2}$$

$$\text{So} \quad D^2 = \frac{(1-A) \, L}{16 \pi \, 5.7 \times 10^{-8} \, T^4}$$

we get the inner radius of the Habitable 'Liquid Water' Zone, and with T= 273 k we get the outer distance of this zone. We can calculate this zone for any star for which we know its luminosity, and are willing to estimate a likely albedo for the orbiting planet.

If the planet has a greenhouse atmosphere, it can trap more of the stellar luminosity and become far hotter than this estimate suggests. For Earth with no atmosphere, its equilibrium temperature is about 255 k or -18 Celsius, rather than the average value of 291 k or 14 Celsius. This difference is caused by the trace amount of carbon dioxide (currently about 390 parts per million). Other planets like Venus are very close to our solar system's HZ, but their 100% carbon dioxide atmosphere makes their surface temperatures close to 460 Celsius! For exoplanets, it will be very important to know how much greenhouse heating is going on. It's not just a matter of checking

Interstellar Travel

their distance to see if it is in the HZ or not. Most estimates of whether a planet is habitable are based on their being in the star's HZ. We cannot check what their atmospheres are like, so many of the current 'Earth analogs' may not actually be cool enough to be habitable. On the other hand, planets calculated to be just outside their star's HZ and too cold to have liquid water may in fact be habitable if they have some greenhouse heating going on! There is also the interesting possibility raised by the moons of Jupiter, which are under tidal stress and have liquid water under their ice-bound crusts. However, it is not clear that interstellar voyagers would be happy, or even technologically capable, of living under those conditions! The bottom line is that we really do want a planet of Earth size, with just a small amount of carbon dioxide to insure liquid water exists on its surface at least for most times of the year.

On the next page we see a graph of all planets near the habitable zones of their stars (green shades). Only those less than 10 Earth masses are labeled. The size of the circles corresponds to the radius of the planets. It is pretty obvious that lots of Jupiter-sized planets have been found in this zone. When you consider that Jupiter and Saturn in our solar system have moons as big as planets, and Titan has its own atmosphere, there may be hundreds of habitable moons orbiting these massive planets!

Habitable Zones

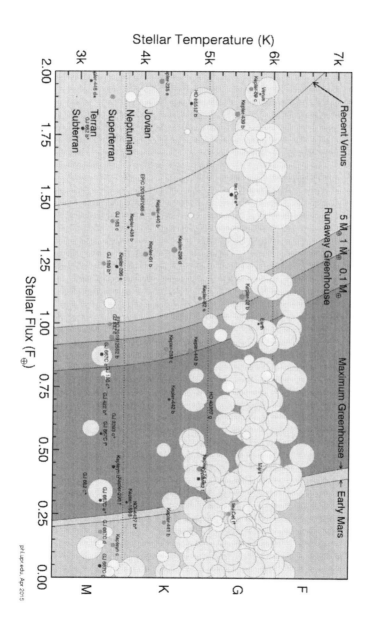

Interstellar Travel

List of destinations…so far!

The List of Destinations…so far!

The 10 nearest stars are: Proxima Centauri (4.24 lys), Alpha Centauri (4.36), Barnard's Star (5.96), Luhman 16 (6.59), Wolf 359 (7.78), Lalande 21185 (8.29), Sirius (8.59), Luyten 726-8 (8.72), Ross 154 (9.68) and Ross 248 (10.32). This takes us out to a distance of just over 10 light years from Earth.

The Alpha Centauri trinary star system has beckoned interstellar travelers for decades with its golden sun-like star Alpha Centauri A, its orange star companion Alpha Centauri B, and the little red dwarf: Proxima Centauri. Life would be so easy were this system to harbor a few interesting planets for explorers to visit. Recent studies hold out diminishing hope that the universe is this kind.

Alpha Centauri B has two verified Earth-sized planets, but not in the star's liquid-water habitable zone. In fact, nature has chosen to be cruel and placed them both in orbits closer than Mercury where they would be uninhabitable due to the extreme heat. This is an artist's impression of the Earth-mass, 1100 kelvin planet Alpha Centauri Bb orbiting every 3 days. Good gravity, but bad location! There are no planets larger than Neptune orbiting this star closer than our planet Jupiter. Meanwhile, nothing has yet been found in orbit around Alpha Centaury A or Proxima. Searches are still in progress to find smaller planets in favorable orbits, but time does seem to be running out for this system. If the *Babylon 5*, Centauri Federation home world exists, it is well-hidden!

Barnard's Star is a red dwarf star barely 3000 k hot and shines with 1/300 the power of our own sun. Why anyone would want

Interstellar Travel

to visit this runt of a star is unfathomable. It was in fact the destination for Project Daedalus - humanities first hypothetical interstellar mission. Astronomer Jieun Choy and his collaborators have intensively studied this star and all available data on it and concluded in 2013 that there is no evidence for planets larger than Earth in orbits with periods from a few hours to 1000 days. This encompasses the entire HZ of this star. *"This nondetection of nearly Earth-mass planets around Barnard's Star is surely unfortunate, as its distance of only 1.8 parsecs would render any Earth-size planets valuable targets for imaging and spectroscopy, as well as compelling destinations for robotic probes by the end of the century."*

Luhman 16 is possibly a worse destination than Barnard's Star. It consists of two brown dwarf stars, each a cool 1300 kelvins and about the size of Jupiter, orbiting about 3 AU apart every 25 years. A possible companion planet orbits one of these stars every month at a distance closer than Mercury. There is so little heat produced by these stars that any planets would be uninhabitable.

As for the stars Wolf 359, Lalande 21185, Sirius, Luyten 726-8, Ross 154 and Ross 248, there have been searches for Jupiter-sized planets around these nearby stars, but none have ever been claimed. Except for Sirius, they are all red dwarf stars whose dim light forces HZ planets to orbit very close to these stars. It is possible that planets may exist in orbits that are face-on to our line-of-sight and have escaped detection so far, so searches continue for any wiggles or changes in the star's motion across the sky as the only way to detect them. Wolf 359 and Sirius have played interesting roles in science fiction, but in reality there

List of destinations…so far!

does not seem to be much there to spawn much of a modern reality-based story!

So, our nearest stars within 10 light years are pretty bleak as destinations for interstellar missions. There is no solid evidence for Earth-sized planets orbiting within the HZs of any of them. These would not be plausible targets because there is so little return on the high cost of getting there, even though that cost in terms of travel time is the smallest of all stars in our neighborhood. Let's take a bigger step and see if we can turn up any new prospects. Out to a distance of 16 light years there are 56 normal stars, which do actually include some promising candidate targets.

Epsilon Eridani (10.52 ly) is a warm, yellow and sun-like star, with one known giant planet orbiting just outside the star's HZ. It also has two asteroid belts: one at about 3 AU and one at about 20 AU. No one would ever risk a priceless mission by sending it to a sparse planetary system with deadly asteroid belts and no HZ candidates! It has been a destination for several hypothetical interstellar voyages by Project Daedalus (1976) and Project Icarus (2011). It is very unlikely that the expense of such missions will be recovered by finding anything of interest that can't be found orbiting these stars from Earth by remote sensing. It has been a frequent subject for science fiction settings, most recently in the 2012 novel *Flight 404* by Simon Petrie. It is also the setting for the *Babylon 5* orbiting space station.

Groombridge 34 (11.62 ly) is another binary star system consisting of two cool red dwarfs orbiting about 147 AU apart.

Interstellar Travel

The only suspected planet has a mass of more than five Earths and orbits the 'A' star every 11 days. No mission would be sent to such a planet for which an atmosphere would probably be crushingly dense and probably Jupiter-like even if it was in its HZ. It appears in the 2013 novel *The Black Hole Project* by C. Sanford Lowe and G. David Nordley, where it is the site of a terraformed colony planet called New Antarctica. It also appears in the *Halo* video game series.

Epsilon Indi (11.82 ly) is a trinary system with a 4,600 k orange star one-quarter as luminous as our sun, and two brown dwarf stars. It appears to have a possible though unconfirmed Jupiter-sized planet with a period of more than 20 years. There are no known smaller planets. It was featured in the 1996 story *Starplex* by Robert Sawyer, and in Larry Niven's 1973 novel *Protector*. The deadly Gorgan from *Star Trek* also originated from this system.

Tau Ceti (11.88 ly) is a sun-like yellow star and probably has five planets between two and six times Earth's mass, with periods from 14 to 640 days. Planet Tau Ceti f is colder than Mars and is at the outer limit to the star's HZ. Its atmosphere might be dense enough for greenhouse heating, so the world might be habitable after all. But this is guesswork not certainty. This star has a huge following in science fiction literature. Virtually every major author has written a story with Tau Ceti as a setting including Arthur C. Clarke, Larry Niven, Robert Heinlein, Isaac Asimov and Frank Herbert. It has been mentioned three times in *Star Trek* stories: *The Wrath of Kahn*, *Conspiracy* and *Where No Man Has Gone Before*.

List of destinations…so far!

Kapteyn's Star (12.77 ly) is another red dwarf star and produces 1/10 the light of our sun. It has two planets, Kapteyn b and Kapteyn c that are 5 to 8 times the mass of Earth. Kapteyn b has a period of 120 days and is a potentially habitable planet estimated to be 11 billion years old. Again, it is a massive planet whose surface you could never visit, so what is the point of the interstellar expedition? If it also has a dense atmosphere given its mass, greenhouse heating might make it a Venus-like world. Again, not a good destination! It hosts a small handful of science fiction stories including the 2003 novel *Fallen Gods* by Jonathan Blum and Kate Orman, and the recent 2014 *Sad Kapteyn* by Alastair Reynolds.

Gliese 876 (15.2 ly) is another cold red dwarf 1/10 as bright as our sun, which seems to have four planets. All are more than 6 times the mass of Earth and orbit closer than the planet Mercury. Gliese 876 c is a giant planet like Jupiter in the star's habitable zone. Would you bet the entire mission that 876c has habitable 'Galilean moons' like our Jupiter? This would be an unacceptable shot in the dark, though a tantalizing one. Not even science fiction writers have warmed up to this system. Only Larry Niven's 2005 novel *Building Harlequin's Moon* uses a moon of the gas giant Harlequin to refuel a starship.

So, out to 15 light years we have some interesting prospects but no confirmed Earth-sized planet in its star's HZ whose surface you could actually visit. We also have no solid data on the atmospheres of any of these worlds. None of these candidates seem worth investing the resources of a trillion-dollar mission to reach and study. We can study them all from Earth at far less cost.

Interstellar Travel

If we take an even bigger step into interstellar space and consider stars closer than 50 light years we have a sample of potentially 2000 stars, but not all of them have been discovered and cataloged. About 130 are bright enough to be seen with the naked eye. The majority are dim red dwarfs, which are still good candidates for planetary systems if these planets are located inside the orbit of Venus to keep their surfaces above the freezing point of water. In this sample we encounter among the known planetary candidates several stars that would be intriguing targets:

61 Virginis (11.41 ly) is a nice, warm yellow star like our sun. It has three planets with masses between 5 and 25 times Earth crowded inside the orbit of Venus. The debris disk that orbits this star has at least 10 times as many comets as our solar system. There are no detected planets more massive than Saturn within 6 AU. An Earth-mass planet in the star's HZ remains a possibility, but the asteroid belts make this an unacceptable high risk target. It has so far never been mentioned in science fiction stories, even though it is one of the most intriguing settings in the vicinity of our sun!

Gliese 667 (23.2 ly) is a trinary system with two sun-sized orange stars and a red dwarf. As many as seven planets may orbit this star, but most have not been confirmed. All have masses between that of Earth and Uranus. All but one are huddled inside the orbit of Mercury. Planets c and d are in the star's HZ and are at least 3 times the mass of Earth. Their hypothetical moons may be habitable. Only Gliese 667Cb and Cc have been confirmed as planets. Again, no science fiction stories about this system...yet.

List of destinations…so far!

55 Cancri (40.3 ly) consists of a sun-like yellow star, and a red dwarf orbiting about 1065 AU from the yellow star. All five detected planets are more than five times the mass of Earth. Only 55 Cancri e is located at the inner edge of the star's HZ and its hypothetical moons could be habitable. More planets are possible within the stable zone between 0.9 to 3.8 AU if their orbits are circular. This is a system we still need to study. Although ignored by science fiction, a METI radio message was sent to 55 Cancri on July 6, 2003, and it will arrive in May 2044. If ET is there, we could get a return reply in the year 2084!

HD 69830 (40.7 ly) is an orange star slightly dimmer than our sun, and is known to have a debris disk produced by an asteroid belt twenty times more massive than the one in our own system.  Three detected planets have masses between 10 to 18 times that of Earth. The debris disk makes this a high-risk prospect to visit and explore even if there are habitable moons. In the *Halo* fictional universe, the Jackal homeworld is the hypothetical moon of the third planet in the system. The planet is a Neptune-like gas giant, and the Jackals (Kig-Yar) evolved on a moon known as "Eayn".

Interstellar Travel

HD 40307 (41.8 ly) is an orange star about ¼ as bright as our sun. Five of the six planets orbit very close to the star inside the orbit of Mercury. The planets range in mass from three to ten times Earth. The fifth planet, HD 40307g, is 8 times the mass of Earth and orbits at a distance similar to Venus in the system's habitable zone. Is a planet like HD 40307g in its star's HZ, with a mass too great for a direct human surface visit a good candidate? I don't think so. Science fiction writers don't think so either.

Upsilon Andromedae (44.25 ly) is a binary star with a red dwarf orbiting a yellow, sun-like star. Of the four known Jupiter-sized planets, the two outer planets are in orbits more elliptical than any of the planets in our solar system. Upsilon Andromedae d is in the system's habitable zone, has three times the mass of Jupiter, with huge temperature swings. Its hypothetical moons may be habitable. The system is also currently untouched by science fiction.

47 Ursa Majoris (45.9 ly) is a virtual twin to our sun only slightly more massive and about twice as old. There are three known Jupiter-sized planets. The innermost world 47 Ursae Majoris b is more than twice the mass of Jupiter and orbits between Mars and Jupiter. The inner part of the habitable zone could host a terrestrial planet in a stable orbit. None yet detected, and they would be expected to be dry, water-poor planets.

So there we have our known list of candidates out to 50 light years. There are still many more stars in this sample to detect, catalog and study, so it is possible that a Goldilocks Planet could

List of destinations...so far!

be found eventually. But we are now looking at candidates more than 20 light years away at a minimum. This will considerably increase the cost and duration of any interstellar mission by factors of five to ten times a simple jaunt to Alpha Centauri.

As we have already seen, over the years science fiction authors have taken a whack at some of these nearby stars as the settings for their stories. Reports of a planet orbiting 61 Cygni by Peter van de Kamp in 1942 served as inspiration for Hal Clement's 1953 science fiction novel *Mission of Gravity*. American physicist Robert L. Forward considered the Epsilon Eridani, system as one of the targets for interstellar travel in 1985. Proposals for interstellar travel have also considered some of these promising based on the information available at the time. In 1986, the British Interplanetary Society suggested Epsilon Eridani as one of several targets in its Project Daedalus paper study. The system has continued to be among the targets of such proposals, as with Project Icarus in 2011.

During Project Ozma in 1960, Epsilon Indi and Tau Ceti were examined for artificial radio signals, but none were found. In 1972, the Copernicus satellite was used to examine this star for the emission of ultraviolet laser signals. Again, the result was negative. Even the SETI program has considered some of these stars favorable targets for their radio searches. Epsilon Eridani was among the target stars of Project Phoenix, a 1995 microwave survey for signals from extraterrestrial intelligence.

Some astronomers have proposed a handy planetary classification scale based on temperature and mass as shown in the graph below. For example, mesoplanets (Class M) would be

Interstellar Travel

ideal for complex life, whereas Class hP or hT planets would only support extremophylic life.

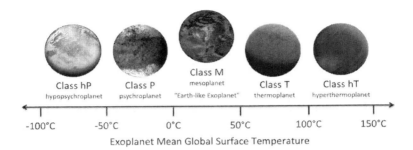

With this classification scheme in mind, the nearest stars with a planet in the habitable zone are currently as follows:

Planet	Distance	Type	Status
Tau Ceti e	11.9 ly	Super-E, Thermo	Unconfirmed
Kapteyn b	12.7	Super-E, Psychro	**Confirmed**
Gleise 832c	16.1	Super-E, Meso	**Confirmed**
Gliese 682c	16.6	Super-E, Psychro	Unconfirmed
Gliese 581d	20.2	Super-E, Meso	Unconfirmed
Gliese 667 Cc	23.6	Earth, Meso	**Confirmed**
Gliese 180b	39.5	Super-E, Meso	Unconfirmed
Gliese 422b	41.3	Super-E, Meso	Unconfirmed
HD 40307g	41.7	Super-E, Psychro	**Confirmed**
Gliese 163c	48.9	Super-E, Meso	**Confirmed**

There are also some purported planets that have a dubious identity or confirmation: Gliese 581g warm superterran mesoplanet and Gliese 581d warm superterran hypopsychroplanet (20.2 ly); and Gliese 667 Cf warm terran psychroplanet and Gliese 667 Ce warm terran psychroplanet

List of destinations…so far!

(23.6 ly). Future studies will clarify whether they are real or just data artifacts.

What could be more exciting than actually seeing a planet with a telescope rather than having to surmise its existence based upon transit or gravitational data? So far there are a total of 15 stars that have had their planets imaged directly using a variety of techniques. All of these planets have to be far enough from their stars that the brilliant starlight can be filtered out, so the planets tend to be more than 20 AU from their stars. The closest of these at 25 light years from our sun is Fomalhaut b, which was imaged by the Hubble between 2004 and 2012.

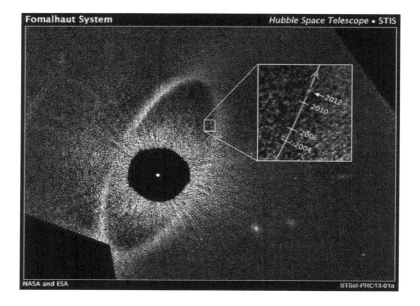

None of these imaged planets were detected in the same way that the thousands of known exoplanets have been, which rely

Interstellar Travel

on astrometric data or direct transits. The ones detected photographically are at least as large as Jupiter.

By far the most impressive planetary system is the four planets orbiting the star HR 8799 located 128 light years from Earth in the constellation Pegasus. It is part of a system that also contains a debris disk. In 2008, astronomers directly observed three planets orbiting the star with the Keck and Gemini telescopes in Hawaii. A search through images by the Hubble Space Telescope also spotted the planets in 1998. Further observations in 2009–2010 revealed the fourth giant planet just less than 15 AU from the star.

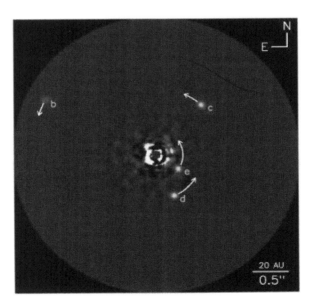

The planets orbit at distances that are two to three times greater than Jupiter, Saturn, Uranus and Neptune, but because the star is five times more luminous than our sun, the planets e, d, c and

List of destinations...so far!

b receive nearly the same amount of heating as Jupiter, Saturn, Uranus and Neptune. Also, the planets are found in a relatively clear region between two debris disks interior to the orbit of e, and farther out than the orbit of b. In fact, the outer disk resembles our Kuiper Belt region and is probably filled with numerous small bodies a few kilometers or thousands of kilometers in size. It is one of the most massive disks known around any star within 300 light years of Earth, and there is room in the inner system for Earth-sized planets. The four planets are at the limit of mass that separates planets from brown dwarf stars. Since the star is only about 30 million years old, the planets are still glowing red-hot due to their young age. Spectroscopic studies suggest that planets b and c appear to have cloudy atmospheres containing molecules of carbon monoxide and methane.

What we are actually looking for are planets that are within their star's habitable zones where liquid water could exist. This turns out to be a rather small collection of known worlds out to several thousand light years (Kepler 62). We cannot discount even massive Jupiter-sized planets because as we know from our own solar system, they can have huge satellite moons that could be a large as Mars or Earth, and so still have habitable conditions.

If you were to plot all these candidate planets on a graph that also shows where they are in their habitable zones, it would look like this figure. Note where Venus, Earth and Mars are located.

Interstellar Travel

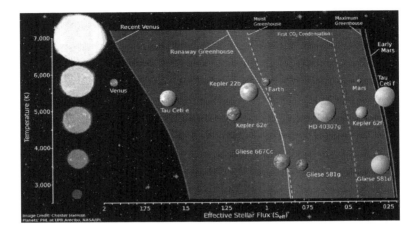

It is a pretty sparse diagram consisting of only the roughly Earth-sized worlds. We can imagine these rocky planets with pools of water on their surfaces, or even entire oceans, and think of them as prime destinations for a future visit. We know from our own solar system that the terrestrial planets have atmospheres that can be as thick as that of Venus and as dilute as that of Mars. But that brings us to another issue. Just because a planet is in its habitable zone and has an atmosphere, that does not make it a nice place to visit. You still have to worry about what its star is doing!

Flare Stars

As seen from a distance, our sun as a star does not produce a constant level of light, especially in the ultraviolet and x-ray portions of the electromagnetic spectrum. The sunspots and solar flares that appear and disappear on an 11-year cycle cause our sun to flare up in brightness and then dim in minutes, hours and years. Distant astronomers would observe these minute brightness changes and classify our sun as a G-type, dwarf Main Sequence flare star. They would also conclude that our sun has sunspots and an 11-year cycles of these spots, with occasional large flares on its surface as magnetic fields get tangled and release their energy.

Most of these events are directed away from Earth and their energy flows harmlessly into the rest of the solar system. But enough are directed towards Earth every year that they are a cause for concern for our astronauts working in space, and for our very expensive collections of commercial and military satellite technology! Because activity cycles similar to our sun are not uncommon, we should assume that the hazardous solar flares and coronal mass ejections we see first-hand are not unique to our sun, though not easily detectable around other stars. Some stars have activity cycles that are faster or slower than our sun's 11-year cycle.

The first flare star observed by astronomers occurred in 1927 in a star known as V1396 Cygni, and later in 1948, UV Ceti and RS Canum Venaticorum were also discovered to change thwir brightness in an irregular but dramatic way. UV Ceti, also called Luyten 726-8, is a nearby star only 8.7 light years from Earth,

Interstellar Travel

and is a binary system consisting of two M-dwarf stars orbiting each other every 26 years. RS Canum Venaticorum is a binary star consisting of F and a K dwarf stars orbiting each other every 24 days, and located 460 light years from Earth. Since then, hundreds of these stars have been discovered.

A flare star normally has a steady output of light, except on occasion it suddenly brightens for minutes or hours before returning to its normal brightness. Based on many studies of this flaring behavior, astronomers think that the flares are caused by enormous magnetic fields shared between the stars in these binary systems that get tangled up and then release their energy. The magnetic energy accelerates particles just the way solar flares are produced, and this causes the light from the star, especially in the ultraviolet and x-ray region, to increase dramatically. In many instances for flare stars like RS Canum Venaticorum, you cannot see the companion, and this has led to the idea that for the flare events near isolated stars, a large Jupiter-sized planet in a very close stellar orbit may be responsible.

Most flare stars, however, are solitary cool red dwarf stars and some brown dwarf stars as well. Flare stars are intrinsically faint, but have been found to distances of 1,000 light years from Earth. The nearby stars Proxima Centauri, Wolf 359 and Barnard's Star are among the closest examples. On April 23, 2014, NASA's Swift satellite detected the strongest, hottest, and longest-lasting sequence of stellar flares ever seen from a nearby red dwarf star. The initial blast was as much as 10,000 times more powerful than the largest solar flare ever recorded.

Flare Stars

Proxima Centauri undergoes random increases in brightness because of magnetic activity. The star's magnetic field is created by convection throughout the stellar body, and the resulting flare activity generates a total X-ray emission similar to that produced by the Sun, which is a larger and much hotter star.

Wolf 359 is a red dwarf of spectral class M6.5 that emits X-rays. It has a relatively high flare rate. The star's magnetic field has an average strength of about 2,200 Gauss, but this varies significantly on time scales as short as six hours. By comparison, the magnetic field of the Sun averages 1 Gauss, although it can be as high as 3,000 Gauss in active sunspot regions.

Barnard's Star is suspected of being a flare star. It is nearly twice as old as our sun and rotates once every 130 days. In 1998, astronomers detected a powerful flare event. This is surprising from such an old star, which was not expected to have a significant magnetic field due to its slow rotation. The flare momentarily raised the apparent temperature of this star from 3,000 k to over 8,000 k. Our sun, by comparison, has a very steady temperature of 5,770 k.

TVLM513-46546 is an ultra-cool brown dwarf located 35 light years from Earth, exhibiting flaring activity. The star has a mass as low as 90 times the mass of Jupiter (or 9 percent of Sun's mass) and it may have an unseen companion planet.

A major problem is that red dwarf stars outnumber sun-like stars by nearly 10 to 1, and their habitable zones are much closer to the star's surface and well inside the orbit of Mercury at the same scale. Significant flaring behavior in these promising

Interstellar Travel

candidates for stars with habitable planets means that visitors will have to worry about sudden radiation exposure events in space. But the more inviting sun-like stars may not be any picnic either

Sun-like stars have habitable zones farther out and near the orbits of our planets Venus, Earth and Mars, but a number of solar-type stars are also known that show flaring behavior. They are often called superflare stars because the energy from such an event is millions of times more intense that the largest flares seen on our own sun. The known superflare stars are Groombridge 1830 (30 ly), Kappa Ceti (29 ly), MT Tauri, π1 Ursae Majoris (47 ly), S Fornacis (480 ly), BD+102783, Omicron Aquilae (62 ly), Serpentis and UU Coronae Borealis. Although the data is different for each star, the energies in the events represents as much energy as the entire sun emits from seconds to days! An Earth-like planet at Earth's orbital distance would temporarily lose its ozone layer, and even as far away as the moons of Jupiter, the daytime side of icy moons would temporarily melt and refreeze.

S Fornacis produced a flare on March 2, 1899 that caused the star to brighten by 16 times in a matter of a few hours. The equivalent power was about 3×10^{27} watts, which is 10 times the power of our own sun! According to some theories, these solitary stars may have nearby Jupiter-sized planets whose magnetic fields interact with the star's magnetic field to produce these monstrous flares.

Although in Larry Niven's 1973 novel *'Inconstant Moon'* a solar superflare devastates the day side of Earth, there is no evidence

Flare Stars

that our sun has ever launched a superflare into the solar system in its billion-year recent history. However, when our sun was just forming as a so-called T-Tauri star, these events were very common.

Young stars tend to have surfaces that are still settling-down from the formation of the star. They also tend to be rotating very fast. Both of these factors make for strong magnetic fields and powerful flares. On April 25, 2008, NASA's Swift satellite detected the flash of light from this nearby star EV Lacertae. It was thousands of times more powerful than the most intense solar flare we have observed so far.

At the distance of a few dozen light years, even the tremendous solar flare on our own sun that occurred on August 2, 1972, and which would have been extremely unhealthy and even fatal for

Interstellar Travel

some astronauts, is not detectable. This means that some of our destination stars may provide us with very unhealthy radiation environments, except on planetary surfaces with a thick atmosphere. Routine operation in space with only the basic cosmic ray shielding we bring with us, could be a hazardous activity. This could be fixed by adding more shielding before we leave to cover worst-case situations, but that means added mass and cost to the mission for shielding we may not have needed.

Future astronomical observations with better technology will be able to measure August 1972-like flares on most stars in the solar neighborhood if they happen every few years, but if they are once-a-century events like the 1859 Carrington Superstorm, we may never have enough time to know this for certain. But beyond superflares, even mostly normal stars can be nasty in their own ways.

Epsilon Eridani has been extensively studied by astronomers for many decades. It is a star several thousand degrees cooler than our sun, and about 80% as massive. With an age of 500 million years, it is among the youngest stars in our solar neighborhood. Epsilon Eridani rotates three times more rapidly than our sun, and has a sunspot cycle of about 5 years. It also varies in brightness by about 5%. The reason for this is that, although most of its surface has virtually no measurable magnetic field, about 9% of its surface is covered by a single large sunspot with a magnetic field of about 200 Gauss. This is a far-weaker field than is found in solar sunspots (typically 3,000 Gauss), but for a cool star like Epsilon Eridani, it is enough to significantly alter its brightness.

Flare Stars

The star also has a higher level of activity than our sun in its corona as well. This leads to the production of x-rays, and a brightness that is much higher than our own sun when our sun is at its most active sunspot state. The x-ray brightness is about 2×10^{21} watts, and it is produced by the corona, which is both larger and hotter than our sun's. Because of its very hot 4 million-degree corona, Epsilon Eridani is losing mass at a rate that is 30 times greater than for our sun, and expands as a dense wind of plasma throughout the surrounding space. Although the solar wind 'heliosphere' from our sun extends about 150 AU from the sun to just beyond the orbit of Pluto, the heliosphere surrounding Epsilon Eridani extends over 8,000 AU across. If we could see it from Earth, even at this stars distance, its heliosphere would appear wider than the full moon.

Other stars within 13 light years that are similar to Epsilon Eridani and are also 'star spot' stars are Barnard's Star, Kapteyn's Star, 61 Cygni, Ross 248, Lacaille 8760, Lalande 21185 and Luyten 726-8.

Groombridge 34 as we saw before, is a binary star system of consisting of two red dwarf stars in a nearly circular orbit with a separation of about 147 AU. Both stars exhibit variability by up to 300% in brightness due to random flares.

So, many of the numerous red dwarf stars in the solar neighborhood that could have habitable zones and planets within them, are also potential problems for solar activity and flares much more vigorous than what we are accustomed to with our own sun. That means we will have to design our radiation shielding carefully, and know a great deal about our

Interstellar Travel

destination star before undertaking the journey. Too much shielding will be an unnecessary cost added to an already expensive mission, but too little will be a severe health hazard.

Kepler recently found three super-Earths orbiting the red dwarf star EPIC 201367065 located 147 light years from Earth. The planets are 2.1, 1.7 and 1.5 times Earth's mass, and the outermost planet, d, orbits at the inner edge of the star's habitable zone. These stars are becoming the favorites for planet searches because the planets must orbit close to the star to be in their habitable zones, and this makes them easier to spot using transit and Doppler techniques. This also means our most common interstellar destinations may be to stars that have unpredictable flares.

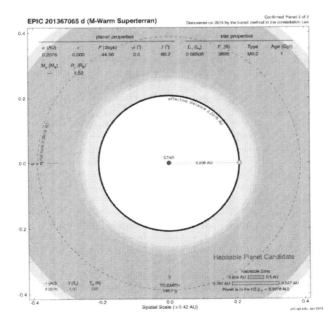

Are we there yet?

Are we there yet?

Most star distances are known to about a tenth of a light year, but that's a whopping 900 billion kilometers. If you want to get somewhere and know a date for arrival, you need to determine star distances to a lot better accuracy. Also, since stars move, you need to point the ship where the star will be in a few centuries, not where it is now.

For stars within 100 light years of the sun, the most accurate way to determine distances is also the most straight-forward. You photograph the star from two different times in Earth's orbit around the sun, and from the slight parallax shift you see between the two photos, you can use a bit of trigonometry to determine distances. The bigger the parallax angle shift, the closer the star.

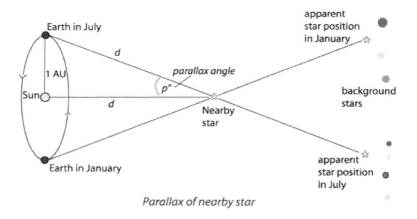

Parallax of nearby star

From a simple proportion, a shift of one arc second in the parallax angle, at a distance of 3.26 light years equals a linear

Interstellar Travel

shift equal to the distance between the Earth and Sun of 150 million kilometers. From this, you can determine the distance to the star, which is just the reciprocal of the parallax angle in arcseconds (1 arcsecond = 1/60 degree; 1 radian=206265 arcseconds). The distance unit is called the parsec and is 3.26 light years long. The European Hipparcos satellite measured the parallax angles for 100,000 nearby stars out to a distance of 1,600 light years. The most accurate distance to the nearest star Proxima Centauri is 1/0.7687 or just 1.3009 parsecs, which equals 4.243 light years. The European Gaia mission launched in 2013 will complete a parallax catalog for 1 billion stars to an accuracy of 10 microarcseconds, leading to very accurate distances for stars as far away as 10,000 light years.

$$\frac{1 \text{ AU}}{1 \text{ parsec}} = \frac{1 \text{ arcsec}}{206265 \text{ arcsecs}}$$

Although the distance accuracy for even a nearby star like Proxima Centauri is measured in thousandths of a light year, since one light year equals 9.5×10^{12} kilometers, this still means that we don't know how far away Proxima is to better than about 9 billion kilometers! The Gaia mission, however, will have an accuracy of 0.000001 arcseconds, which means a star as far away as 50 light years (15.3 parsecs or 4.7×10^{14} kilometers) will have its position determined to about 2,400 kilometers!

$$\frac{X}{15.3 \text{ parsec}} = \frac{0.000001 \text{ arcsec}}{206265 \text{ arcsecs}}$$

$X = (4.7 \times 10^{14} \text{ km}) \times (1 \times 10^{-6} / 2 \times 10^{5})$

$X = 2,400$ kilometers

Are we there yet?

This all sounds very good for determining an accurate arrival time to better than a few minutes, but there is a second problem. All stars move relative to our sun, and by the time you arrive at the star, it will have moved far from where you saw it when you left Earth.

Astronomers measure the three components of a stars velocity by using photographs and spectrometers. By taking pictures of the same star separated by many years, you can measure the star's 2-D motion across the sky. This is called the star's proper motion and is measured in arcseconds of angle per year. You can convert this angle into a distance by knowing the star's actual distance. To get the third dimension of the motion, you have to study the light from the star.

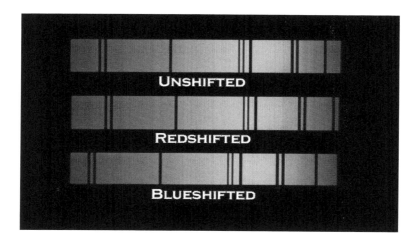

This figure shows the simulated light from a star seen with a spectroscope. The black lines are due to specific elements like hydrogen and calcium. The top spectrum is what you would see if the object is not moving. The middle shows the object

Interstellar Travel

moving away from you at constant speed, while the bottom spectrum shows the object moving towards you at constant speed. This is the Doppler Effect, and by measuring how much the lines shift, you can figure out exactly how fast the object is moving towards or away from you. By also using the Doppler Shift of the star's light, you can measure its motion in the third space direction along our line-of-sight to the star. This is measured in kilometers per second. So, from simple measurements made at Earth you can figure out, not only how far away a star is by its parallax shift, you can also figure out the three space components to its velocity through space: two in the plane of the sky (called the transverse velocity), and the third along the line-of-sight (called the radial velocity). With this information, you can now accurately predict exactly where the star will be when you leave Earth, and where you should 'point your rocket' when you leave so that you arrive at the destination.

$$\text{Speed} = C \times \frac{\text{Difference}}{\text{Wavelength}}$$

C = speed of light

Wavelength of line at rest = 6000 Angstroms

Wavelength of line in star = 6010 Angstroms

$$\text{Speed} = 300000 \times \frac{10}{6010}$$

Speed = 500 km/sec

As we have seen, parallax measurements are accurate to a few thousand kilometers or better out to 50 light years. The proper motion and Doppler shift measurements with the help of the Gaia mission will have accuracies better than 1 km/sec.

Are we there yet?

For example, this image shows the proper motion across the sky of Barnard's Star between 1991 and 2007. It moves -0.8 arcseconds/year east-west and 10.3 arcseconds per year north-south for a total sky motion of 10.4 arcseconds per year. At its distance of 6 light years, this 2-D sky motion is about 92 km/sec. Its line-of-sight 'radial velocity' speed has been Doppler measured as 111 km/sec, so its total speed in 3-D is 144 km/sec.

$$V^2 = 92^2 + 111^2$$
$$V^2 = 8464 + 12321$$
$$V^2 = 20785$$
$$V = 144 \text{ km/sec}$$

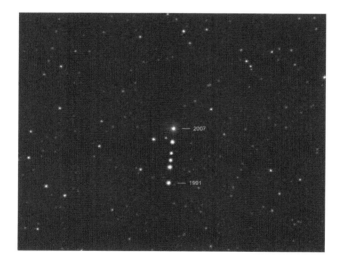

Knowing the relative speeds of nearby stars to 1 km/sec accuracy means is that for our estimated journey of 250 years, we will be able to predict the new position of the star to about D = 1 km/sec x 250 years = 8 billion kilometers. That is the distance from the sun to just beyond Pluto (6 billion km). For the closest stars, even higher accuracy is possible so that there is little to prevent the travelers from arriving within a few million

Interstellar Travel

kilometers of their planned destinations. At the interplanetary speeds of tens of km/sec they will have decelerated to by then, this distance uncertainty translates into an ETA error of about two days, after a journey of possibly hundreds of years!

A bit of relativity!

There is another thing to consider too. At the expected speed of 20% the speed of light, we get a little help from special relativity. Clocks in moving objects run more slowly than the clocks left behind on Earth, so although our friends on Earth will see our journey take 250 years, at this speed the ship's clocks will say that only about 245 years will have elapsed for the travelers!

$$\text{Earth time} = \frac{\text{Ships Time}}{[1 - B^2]^{1/2}}$$

Earth time = 250 years

B = 0.2 of light speed

Ships Time = $250 \times [1-.2^2]^{1/2}$

Ships Time = 245 years.

This little graph will tell you the one-way travel time in ship's years, including relativity, for trips at various speeds given in fractions of the speed of light. For example, the fastest spacecraft speed so far was reached by Voyager 2 at 17 km/sec. That's about 0.00006 of light-speed, and will take about 360,000 years to travel 20 light years! Our zippy interstellar ship could do this in about 100 years at 0.2 times light-speed.

Are we there yet?

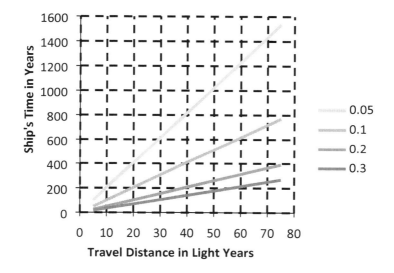

Interstellar Travel

Exoplanet Atmospheres

Humans are delicate creatures. Not only are we hemmed in by gravity considerations, but also by what we breathe.

Oxygen. Oxygen is essential to human life. The human body takes the oxygen breathed in from the lungs and transports it to the other parts of the body on the red blood cells. Oxygen is used and required by each cell. Most of the time, the air in the atmosphere has the proper amount of oxygen for safe breathing. However, the level of oxygen can drop due to other toxic gases reacting with it. The minimum oxygen concentration for human breathing is 19.5 percent. For humans and many animals to sustain normal functions, the percentage of oxygen in the breathing environment must be within a small range.

The Occupational Safety and Health Administration, OSHA, determined the optimal breathing range to be between 19.5 and 23.5 percent oxygen. Serious side effects will occur if the oxygen levels are outside of the safe zone. At levels at or below 17 percent, your mental abilities become impaired. When levels drop to 16 percent or below, noticeable changes begin while levels under 14 percent will cause extreme exhaustion from physical activity. Once levels drop below 10 percent, you may become very nauseous or lose consciousness. Humans won't survive with levels at 6 percent or lower. With extremely high concentrations of oxygen, humans can experience harmful side effects. Very high levels of oxygen cause oxidizing free radicals to form. These free radicals will attack the tissues and cells of the body and cause muscle twitching. The effects from short

Interstellar Travel

exposure can most likely be reversed, however lengthy exposure can cause death.

Nitrogen gas kills you, not because it's toxic to your internal chemistry, but because it is simply not oxygen.

Carbon Dioxide Your body, of course, has to get rid of CO_2. The higher the concentration of atmospheric CO_2, the harder it is to do so. CO_2 does bind to hemoglobin, and while it is a different site than where oxygen binds, it does reduce the oxygen carrying capacity of the blood. The higher the CO_2 concentration, the more acidic the blood. The body produces CO_2 and it has a system for monitoring how much CO_2 there is in the blood. If that level is high, it increases the breathing rate to get rid of the excess. In an atmosphere of pure CO_2 that would tend to drive residual oxygen out of the body faster than an atmosphere of, for example, nitrogen. However there's another mechanism that's more likely to make a difference. If the brain is deprived of oxygen you faint. Normally, that's a good thing since it makes you lie down and that stops gravity draining blood and oxygen away from the brain. However, if you are in a room with no oxygen and you faint then you are not in a position to save yourself. Carbon dioxide is heavier than air and clings to the ground in a layer.

Methane is odorless and not toxic, yet it is extremely flammable and may form explosive mixtures with air. Methane is violently reactive with oxidizers, halogen, and some halogen-containing compounds. Methane is also an asphyxiant and may displace oxygen in an enclosed space. Methane is extremely flammable and can explode at concentrations between 5% (lower explosive

Exoplanet Atmospheres

limit) and 15% (upper explosive limit). These concentrations are much lower than the concentrations at which asphyxiant risk is significant. Reportedly, the most violent methane explosions occur at concentrations of about 9%; coal mines are hence kept well ventilated (pumped with fresh air) to maintain methane levels at or below 1%. Methane gas is lighter than air and the highest concentrations inside buildings are found near the ceiling.

Ammonia gas – Ammonia is highly toxic. Normally blood ammonium concentration is < 50 micromoles per liter (μmol /L), and an increase to only 100 μmol /L can lead to disturbance of consciousness. A blood ammonium concentration of 200 μmol /L is associated with coma and convulsions. Ammonia is an irritant and irritation increases with concentration; the Permissible Exposure Limit is 25 ppm, and lethal above 500 ppm.

Carbon monoxide - Carbon monoxide (CO) is produced by incomplete combustion of carbonaceous material. Commonly overlooked or misdiagnosed, CO intoxication often presents a difficult for people to detect. Exposures at 100 ppm or greater can be dangerous to human health. Carbon monoxide exposure may lead to a significantly shorter life span due to heart damage.

Hydrogen – (H2) - Hydrogen isn't poisonous, but if you should breathe pure hydrogen you could die of asphyxiation simply because you'll be deprived of oxygen. Worse, you won't necessarily know that you're breathing hydrogen because it's invisible, odorless and flavorless -- much like oxygen. Another concern is that hydrogen flames are nearly invisible. When

Interstellar Travel

hydrogen catches fire, the flames are so dim and hard to see that they're both hard to avoid and hard to fight.

Argon - Although argon is non-toxic it is considered a dangerous asphyxiant in closed areas. It is also difficult to detect because it is colorless, odorless, and tasteless.

Table 2. Composition of the Atmospheres of Jupiter, Saturn, Uranus, and Neptune				
Gas	Jupiter[a]	Saturn	Uranus	Neptune
H_2	86.4 ± 0.3%	88 ± 2%	~82.5 ± 3.3%	~80 ± 3.2 %
^4He	13.6 ± 0.3%	12 ± 2%	15.2 ± 3.3 %	19.0 ± 3.2 %
CH_4	(1.81 ± 0.34) × 10^{-3}	(4.7 ± 0.2) × 10^{-3}	~2.3 %	~1-2 %
NH_3	(6.1 ± 2.8) × 10^{-4}	(1.6 ± 1.1) × 10^{-4}	<100 ppb	<600 ppb
H_2O	520^{+340}_{-240} ppm	2–20 ppb		
H_2S	67 ± 4 ppm	<0.4 ppm	<0.8 ppm	<3 ppm
HD	45 ± 12 ppm	110 ± 58 ppm	~148 ppm	~192 ppm
$^{13}CH_4$	19 ± 1 ppm	51 ± 2 ppm		
C_2H_6	5.8 ± 1.5 ppm	7.0 ± 1.5 ppm		
PH_3	1.1 ± 0.4 ppm	4.5 ± 1.4 ppm		
CH_3D	0.20 ± 0.04 ppm	0.30 ± 0.02 ppm	~8.3 ppm	~12 ppm
C_2H_2	0.11 ± 0.03 ppm	0.30 ± 0.10 ppm	~10 ppb	60^{+140}_{-40} ppb
HCN	60 ± 10 ppb	<4 ppb	<15 ppb	0.3 ± 0.15 ppb
HC_3N			<0.8 ppb	<0.4 ppb
C_2H_4	7 ± 3 ppb	~0.2 ppb[b]		
CO_2	5-35 ppb	0.3 ppb	40 ± 5 ppt	
C_2H_6			10 ± 1 ppb	$1.5^{+2.5}_{-0.5}$ ppm
CH_3C_2H	2.5^{+2}_{-1} ppb	0.6 ppb	0.25 ± 0.03 ppb	
CO	1.6 ± 0.3 ppb	1.4 ± 0.7 ppb	<40 ppb	0.65 ± 0.35 ppm
CH_3CN				<5 ppb
GeH_4	$0.7^{+0.4}_{-0.2}$ ppb	0.4 ± 0.4 ppb		
C_4H_2	0.3 ± 0.2 ppb	0.09 ppb	0.16 ± 0.02 ppb	
AsH_3	0.22 ± 0.11 ppb	2.1 ± 1.3 ppb		

[a] ^{3}He 22.6±0.7 ppm, Ne 21±3 ppm, Ar 16±3 ppm, Kr 8±1 ppb, Xe 0.8±0.1 ppb.
[b] assuming a total stratospheric column density of 1.54%×10^{25} cm^{-2}.
From Lodders & Fegley 1998 and updates: Mahaffy et al. 2000, Atreya et al. 2003, Lodders 2004, Wong et al. 2004

Exoplanet Atmospheres

Nitrogen dioxide – Nitrogen dioxide is toxic by inhalation. However, as the compound is acrid and easily detectable by smell at low concentrations, inhalation exposure can generally be avoided. Symptoms of poisoning tend to appear several hours after inhalation of a low but potentially fatal dose. Also, low concentrations (4 ppm) will anesthetize the nose, thus creating a potential for overexposure.

Hydrogen cyanide (HCN) – A hydrogen cyanide concentration of 300 ppm in air will kill a human within 10–60 minutes. A hydrogen cyanide concentration of 3500 ppm will kill a human in about 1 minute. The toxicity is caused by the cyanide ion, which halts cellular respiration.

Pressure

The Earth's atmosphere contains 20.9% oxygen, but the critical factor is its partial pressure. The total air pressure at sea level is 14 pounds per square inch or 101,000 Pascals. This is composed of the 'partial pressures' of the constituents to air, namely nitrogen and oxygen. Nitrogen contributes 79,000 Pascals, oxygen 21,000 Pascals and Argon and carbon dioxide about 2,000 Pascals total. These partial pressures all add up to a total pressure of 101,000 Pascals at sea level.

Oxygen moves from the air to the blood because the partial pressure of air is larger than in the blood. Carbon dioxide moves the other way for the same reason. The real issue with lower partial oxygen pressures is that the human body cannot absorb enough oxygen. At oxygen partial pressures less than 33% of

Interstellar Travel

normal (equivalent to ~9,000 meters or 7,000 Pascals) very few humans would be able to survive.

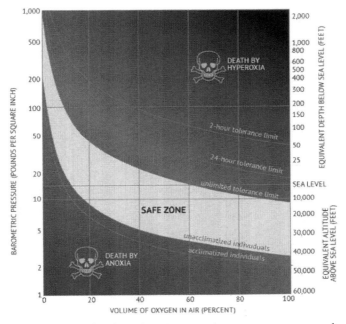

The presence of other (inert) gases is not necessary as long as oxygen partial pressure is at least ~50% of normal or 10,000 Pascals. At about 12,000 meters (40,000 feet) above sea level, the pressure is low enough that most humans will get hypoxia while breathing pure oxygen. This sets a pressure limit of about 18,800 Pascals. At human body temperature (about 37C), water will boil at about 6,330 Pascals. Probably, at a pressure of about 10,000 Pascals, even in a pure oxygen atmosphere, most people would die from anoxia, well above the ambient pressure where bodily water to start to boil. Total atmospheric pressure at the

Exoplanet Atmospheres

summit of Mt. Everest is about 34,000 Pascals. Air is about 20% oxygen, so the partial pressure for O_2 on Mt. Everest is about 7,000 Pascals. The average population probably needs about 14,000 Pascals partial pressure O_2, minimum. On Skylab, total pressure is also about 34,000 Pascals, with 70% oxygen, 30% nitrogen. So, you can survive with a reduced total pressure so long as you have enough O_2 to provide 14,000 to 20,000 Pascals of pressure so your circulatory system works. This high oxygen percentage is, of course, a major fire hazard!

When it comes to finding a good planet with a healthy atmosphere for humans, it all comes down to spectroscopy! Astronomers can figure out the chemical composition of distant objects in the universe like stars and nebulae because we can read the atomic fingerprints of elements and compounds in the light from these distant objects. It is one of the 'miracles' of science and technology first used in the 1800s, and has become the central technique of astronomers for over 150 years.

There are two ways we can use this technique to study the atmospheres of distant planets long before we make the trip to them for a first-hand investigation. We can gather the very faint light from a planet and study its spectrum directly as it emits the light from specific elements, or we can wait until it passes across the face of its star and study the spectrum of its light as its atmosphere absorbs light from specific elements. Either way, we get the same atomic information.

The challenge with today's technology is that planets are so dim that measuring their light and spectrum directly cannot yet be done. But the 'transit' method is getting easier to do every year.

Interstellar Travel

During a transit, the intense light from a star passes through the atmosphere of a planet. The molecule of, say, carbon dioxide absorbs some of this light at specific wavelengths and re-emits it in many different directions. Because most of this star light is now no longer headed in our direction, we see it as a slight dimming in the brightness of the starlight in our direction. Astronomers call this an absorption line spectrum. With modern technology, which consists of very big telescopes to gather the

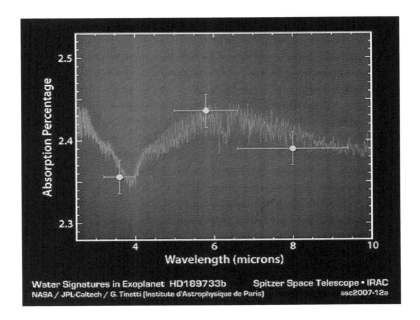

faint light and sensitive spectroscopes, we can detect these absorption lines caused by compounds in the planetary atmosphere. We can only do this, however, for those planets that actually transit their stars as viewed from Earth. The stars

Exoplanet Atmospheres

also have to be bright enough so that enough light is available. This figure shows the predicted carbon dioxide absorption by the planet, and three data points measured by spectroscopy.

Currently, all of the exoplanets discovered by the Kepler mission have the right 'transit geometry' for this method to work, but with current technology the stars are simply too faint for current technology to gather enough light to reveal the tell-tail absorption by exoplanet atmospheres. But, this method has been tried on many other nearer and brighter stars to give astronomers their first glimpses of what these atmospheres might look like.

Planet Name	Distance	Mass	Composition
HD209458b	154 ly	HJ	H2O, CO2, CH4, H, Na
HD189733b	63 ly	HJ	H2O, CO2, CH4, H, Na
XO-1b	560 ly	HJ	H2O
WASP-17b	1000 ly	HJ	Na, H2O
XO-2b	483 ly	HJ	K, Na
2M1207b	172 ly	HJ	H2O, CO
HR8799c	129 ly	CJ	CO, CH4
Gleise 436b	33 ly	HN	CO
Gleise 1214b	42 ly	SE	H2O?
Tau Bootis b	51 ly	HJ	CO, H2O
HAT-P-11b	120 ly	HN	H, H2O
WASP-12b	871 ly	HJ	C, H2O
WASP-19b	815 ly	HJ	H2O

HJ=hot Jupiter, CJ=cold Jupiter, HN = hot Neptune, SE=superEarth.

Because the detection and study of exoplanet atmospheres is so crucial to finding habitable planets, many new missions will be devoting their time to studying nearby exoplanets to figure out

Interstellar Travel

their chemistry. HD 209458b orbits so close to its star it is actually evaporating before our eyes, sheading a comet-like tail of mostly hydrogen gas!

NASA's James Webb Space Telescope (JWST) to be launched in 2018 will measure the infrared light from thousands of exoplanets to detect important molecules in their atmospheres. Many of these stars will be the candidates detected by the TESS mission so that enough light is available to study the planetary atmospheres in detail.

NASA's Transiting Exoplanet Survey Satellite (TESS) mission will be a Kepler-like spacecraft that will monitor the light from 500,000 of the brightest stars in the sky. Scheduled for launch in 2017, it is designed as the first all-sky survey. TESS would spend two years of an overall three-year science mission searching both hemispheres of the sky for nearby exoplanets. *"This is an incredibly exciting time for the search of planets outside our solar system,"* said Mark Sistilli, the TESS program executive from NASA Headquarters in Washington, D.C. *"We got the green light to start building what is going to be a spacecraft that could change what we think we know about exoplanets. During its first two years in orbit, the TESS spacecraft will concentrate its gaze on several hundred thousand specially chosen stars, looking for small dips in their light caused by orbiting planets passing between their host star and us,"* said TESS principal investigator George Ricker of the Massachusetts Institute of Technology in Cambridge, Mass.

TESS is expected to find more than 5,000 exoplanet candidates, including 50 Earth-sized planets. It will also find a wide array of exoplanet types, ranging from small, rocky planets to gas giants.

Exoplanet Atmospheres

Some of these planets could be the right sizes, and orbit at the correct distances from their stars, to potentially support life.

Meanwhile, the United Kingdom's Twinkle mission will be launched in the early-2020s and study the light from 100 exoplanets in the infrared, where molecular signatures are easier to see.

These missions only take us out to the mid-2020s, and many more similar missions will come in the decades after that. The search for an Earth-like planet with a welcoming atmosphere and perhaps even bio-signature gases like oxygen has only begun!

There is one other important factor about planetary atmospheres that can completely change the game on habitability. Greenhouse gases such as carbon dioxide, water vapor and methane even in trace quantities can increase a planet's temperature way above the boiling point of water. The only reason our Earth is habitable in the 'habitable zone', is because it has about 390 parts per million (0.039 percent) of carbon dioxide in its atmosphere. This is all that is needed to make sure we are not a permanently-frozen world at -18 C everywhere. We have an active crust and volcanism that adds CO_2 to the atmosphere, and a water cycle that buries carbon dioxide in carbonate minerals in the ocean water. But too much carbon dioxide like Venus would heat the planet to over 700 F. So our exoplanet atmospheres have to toe a very narrow line in terms of volcanic activity and surface water to insure that the planet is not an ice house or an inferno for life. The problem is that water, carbon dioxide and methane are three of the most

Interstellar Travel

abundant gases in the universe. Entire interstellar gas clouds are filthy with these compounds (CO which is found everywhere is related to CO_2). Statistically, we expect to find many more planets that are perfect in all other aspects, to fail the requirement that they have only trace amounts of these gases.

The bottom line is that, **no** matter what your mode of travel, we will NOT make the journey unless we know well in advance that there is an Earth-massed planet there, in the star's habitable zone where you can have liquid water, and with an atmosphere that is breathable. At the current rate of exoplanet detection and study, we will have a catalog of all Earth-massed planets out to at least 1000 light years from Earth by the end of this century...I promise! Statistically, 25 percent of these planets will be in the star's habitable zone.

But the kicker is the atmosphere. After mounting a multi-trillion-dollar mission, and traveling for centuries to get there, why in the world would you settle for living in a sealed colony on the surface of such a world with a non-breathable atmosphere? Instead you might want to continue a search from Earth for a water-rich planet like Earth with an oxygen atmosphere. Oxygen atmospheres are only produced by biospheres, so what you are looking for is a planet you already know has life...and bacteria and viruses lethal to humans. Now the search for our interstellar destination gets wrapped up in whether you think that after searching the millions of stars up to 1000 light years from Earth, that we will have found our first living planet...or do you think life is even rarer?

Exoplanet Atmospheres

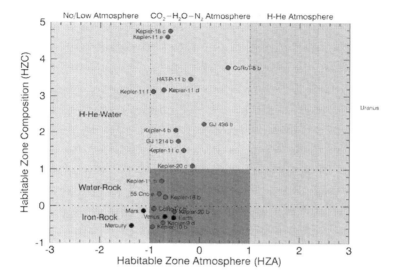

We can at least look at the dynamics of atmospheres and narrow our searches to planets large and cool enough to have atmospheres in the first place!

This graph shows predicted exoplanet atmosphere compositions based on their sizes and distances from their stars. This determines the temperature of their atmospheres. Planets in the darker zone are likely to have stable atmospheres with these compositions. Planets in inner zone are too small to hold onto atmospheres based on heavy molecules like carbon dioxide and water. Planets in outer zone are massive and cold enough they can even hold on to fast-moving hydrogen and helium atmospheres.

Interstellar Travel

A Matter of Extreme Gravity

NASA and military researchers have made strides in figuring out what the acceleration limits for humans are. There are two distinct types: Lateral and Longitudinal. Lateral acceleration is the side to side motion you get in a car taking a sharp curve. At 14-Gs of lateral acceleration, your internal organs will be torn loose from each other during the jolt and this is not survivable. Longitudinal 'head-to-foot' motion, meanwhile, plunges all the blood to the feet. Between 4 and 8 longitudinal Gs will knock you out. Forward or backward accelerations appear to go easiest on the body because they allow the head and heart to accelerate together. Military experiments in the 1940s and 1950s with a rocket sled suggest we can slow down front-to-back at a rate of 45-Gs and still live to talk about it.

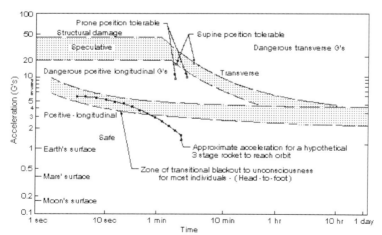

The diagram above shows the human acceleration tolerance for various Gs and times. For sustained accelerations more than one hour, the limit seems to be about 3-Gs. Through the

Interstellar Travel

combination of special G-suits and efforts to strain muscles—both of which act to force blood back into the brain—modern pilots can typically handle a sustained 9*Gs*. According to studies at NASA's Ames Research Center, a person has survived twice Earth's gravity for 24 straight hours without ill effects. However, at four times Earth's gravity or above, human physiology cannot maintain sufficient blood-flow to the brain. In the future, it may be possible to break the 4-G limit with enhancements in genetic manipulation and extremely strong mechanical replacement organs to keep our body's systems running, but that technology is a long way off.

Among the potentially habitable planets now discovered, those in the top row of this figure are the most Earth-like in mass and temperature ranges, however, even these are significantly more massive than Earth implying higher surface gravities. For rocky planets of constant density, the mass varies as R^3 while from the law of gravity, the surface acceleration varies as M/R^2, so the acceleration varies as R^3/R^2 or R, which is a linear increase with the radius of the planet. Rocky planets twice the diameter of Earth have twice the surface gravity of Earth. That means virtually all the exoplanets in the figure that are below the first row in the figure have surface gravities at least twice Earth-normal of 32 feet/sec^2 or 9.8 m/sec^2. You will feel at least twice as heavy as you feel on Earth, and this will have long-term consequences for a human physiology tuned by evolution to live on a 1-G planet.

A Matter of Extreme Gravity

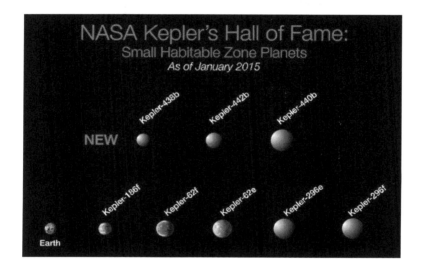

This figure above shows some of the best Earth-sized candidates that have been detected in their habitable zones where liquid water could exist, at least theoretically. Only Kepler 438b at 470 light years and Kepler 186f at 492 light years have virtually the same surface gravity as Earth. The good news is that this is only from a survey of about 4000 stars out of literally millions of stars out to the same distance limits. Closer examples will surely be found.

Interstellar Travel

Are we missing something?

Are we missing something?

No methods for finding and studying exoplanets are perfect, but hopefully with many different methods being used, the chance that some spectacular planet will fall between the cracks becomes small. There are six major methods in use today that have already proven their mettle.

Astrometry searches for the change in position of a star over time as the gravitational force of an unseen planet pulls the star slightly back and forth as it orbits. This requires an accurate telescopic measurement of the star's position in the sky over the course of many years. The first established proof of this method was in 2002 with the Hubble Space Telescope detecting a planet orbiting the star Gliese 876. The new ESA *Gaia* orbiting telescope will use this method on one billion stars and is expected to find many planet candidates.

Spectroscopic methods search for the change in speed of a star caused by a planet's gravitational tugs. This requires obtaining a spectrum of the star over the course of many years and measuring the Doppler shift speed of the star to an accuracy of meters per second. This 'radial velocity' method is used by HARPS (High Accuracy Radial Velocity Planet Searcher) spectrometer at the ESO 3.6-meter telescope in La Silla Observatory, Chile, or the HIRES spectrometer at the Keck telescopes.

Photometry is the measurement of a star's brightness from Earth. When a planet passes across the face of its star, the light from the star dims slightly by a percent or less. By measuring

Interstellar Travel

this dimming from repeated transits you can determine the diameter of the planet, its orbit period, its distance from the star, and by using spectroscopy you can measure the composition of its atmosphere. This is the method famously used by NASA's Kepler observatory.

Gravitational microlensing takes advantage of Einstein's theory of general relativity and the fact that gravity can act like a lens to focus and amplify light. When a planet is between the star and an observer on Earth, the light from the star can brighten very slightly. This is the opposite of the transit method, which measures the dimming of the star's light by an eclipsing planet. This method is the basis for OGLE (the Optical Gravitational Lensing Experiment).

Pulsars spin at a precise rate that can be measured by radio telescopes to many decimal places. They act like super-precise clocks. As a planet orbits a pulsar, the measured pulse arrival times change because of the minute gravitational tugs on the pulsar. This method was used in 1992 to detect several small planets orbiting the pulsar PSR 1257+12.

Direct imaging is exactly what it says. A telescope can actually detect the light from a planet if it is far enough away from its star that you can block out the blinding star light by making an artificial eclipse with an instrument called a coronagraph. This method has been used by many telescopes including the Hubble Space Telescope, the Hale Telescope, and the ESA Very Large Telescope with great success.

Are we missing something?

Transit methods, Doppler methods and gravitational lensing methods all rely on the fact that the orbit plane of the planet is tilted so that it passes very close to the line-of-sight between Earth and the star. If the tilt is too high, the planet will never pass across the face of the star as viewed from Earth and you will have no clue that the planet exists. A detailed study of detection probabilities by Thomas Beatty and Sara Seager at Ohio State University and MIT show that 4% of all stars will have planets detectable this way if the orbits are exactly edge-on, (astronomers call this an orbit inclination angle of 90 degrees) but that this falls to about 1% if the orbits are tilted 10 degrees to the line-of-sight (an orbit inclination angle of 80 degrees).

The Kepler study of 157,000 stars bears this out. About 4000 planet transit candidates were discovered, which is a yield of about 3%. If the orbits are evenly distributed in angle between inclinations of zero and 90 degrees, we get about 1700 per degree or 1% per degree. We should expect to find 1% of our planets with face-on orbits (inclination of 90 degrees) and 1% with inclinations of 45 degrees, and so on. But this also means that transit methods miss about 80 out of 90 planets that are not in the 80-90 degree inclination range for edge-on viewing. Among these could be planets that are even more Earth-like than the ones we detect with the transit method.

The situation is not much better for the Doppler radial velocity or gravitational microlensing methods, which also require this precise alignment. The advantage of the Doppler method is that, although it cannot tell you the diameter of the planet directly, it can tell you the total mass of the planet because gravity depends only on the mass of the planet. The caveat is that if the orbit is

Interstellar Travel

too inclined, the speed change decreases with the sin of the inclination angle so that for a face-on orbit, the change in the star's speed is essentially zero and the planet becomes undetectable. This 'Vsin(i)' effect means that a small planet in an edge-on orbit can look like a massive planet in a much farther orbit that is nearly face-on to the observer at Earth. In fact, all the planet masses we know from the Doppler method are actually minimum values because the observed mass equals the true mass multiplied by the sine of the inclination angle, which cannot be determined from the Doppler method. Although you can estimate a mass for the planet, you cannot determine its diameter or learn anything about the composition of its atmosphere unless you also observe them transit their star and dim the starlight.

Direct imaging works real well for planets in any kind of orbit so long as the planet is far away from its star so that the starlight can be blocked. But this means the method can't really be used right now to see planets closer that Jupiter in our solar system, though we can see these planets much further from their stars. You cannot, however, determine the mass of the planet this way. You can, however, study the atmospheres of these planets at least in principle. Also, you can observe them in orbits with all possible inclinations, which means if you combine this method with the Doppler method, you might determine the planets mass if you can directly image a Doppler candidate.

The exoplanet discovery space is represented in this "orbit size-vs-planet mass" diagram. The solar system planets are indicated by their letters. This plot was produced using online data from *The Extrasolar Planets Encyclopaedia*. Here's how to read it!

Are we missing something?

The graph is logarithmic along both axis in 'powers of ten'. Vertically it indicates the planet mass in multiples of our Earth's mass so '10' means 10 times the mass of Earth. The horizontal axis is the distance from the planet to its star in Astronomical Units where 1 'AU' is the distance from Earth to the sun. The top red band means objects that are more than 13 times the mass of Jupiter (4,000 times the mass of Earth) These objects are classified as brown dwarf stars and are not planets. The smallest known brown dwarf has a mass of 13 times Jupiter and is just barely able to fuse deuterium as a nuclear fuel. The lower red band is objects smaller in mass than our moon.

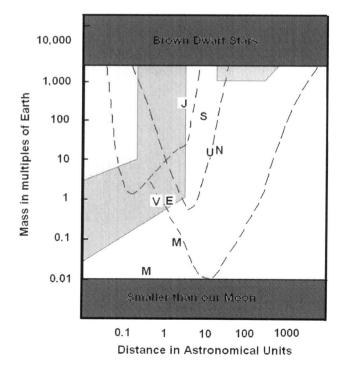

Now things

Interstellar Travel

get complicated.

The shaded areas in represents the sizes and distances of planets that we can detect from ground-based telescopes as they transit their stars. The ground-based searches are shaded on the far-left. The additional objects that the Kepler telescope can see from Earth orbit as they transit their stars are shaded to the right of the ground-based searches. The small shaded area at the top represents objects far enough from their stars that we can directly photograph them. You can see that these objects currently are pretty large and have to be located farther than Jupiter in order for them to be detected with current technology, but in time this region will expand to the left and downwards.

The three dashed lines show the lower limits for detecting planets with microlensing from space (lower curve), microlensing from ground-based telescopes (middle curve) and using the popular Doppler methods (top curve). The Doppler method is good at detecting objects more massive than Earth out to about 10 AU. The gravitational microlensing methods, especially from space, hold enormous promise for finding objects as far away as 100 AU, as well as more massive planets even farther away.

I did not include astrometry in this graph because there are hardly any candidates that have actually been found this way...yet. The ESA Gaia observatory, launched in 201, promises to change that situation dramatically. It is a telescope with a 1.5- mirror, equipped with a gigapixel CCD camera, which will image more than a billion stars at least 70 times over its 5 year mission. This will provide astronomers with the most accurate catalog of

Are we missing something?

stars in this corner of the Milky Way. Gaia's ability to measure star positions in the sky to 20 millionths of an arcsecond (that's 0.000000006 degrees!), means it can see wiggles in the motion of stars from Jupiter-sized planets orbiting anywhere within 10 AU of the star. The 5-year mission length also gives planets closer than 10 AU enough time to be observed as they orbit more than once. It is expected that thousands of new planets will be detected in this way.

So in the next few decades, we will explore all of the bright stars in the sky, and most of the nearby ones using many different techniques that have their own special gaps, but taken together it will be harder and harder for interesting planets to avoid detection. The easiest task is to determine whether a planet exists in orbit around the star. The next-hardest task is to determine it mass. Finally, the hardest thing to do is to figure out if it has an atmosphere and what its composition might be.

Every method has its good and bad points when it comes to finding that elusive Earth-sized planet with a life-supporting atmosphere. Transit methods can give you the orbit size and a good estimate for the mass of the planet, but only the nearest ones provide enough light to study the atmospheres spectroscopically. Direct imaging would let you study the atmosphere of a planet directly, but these planets are bigger than Jupiter and are located outside the habitable zones of their stars so they don't tell us much that is useful. Gravitational microlensing tells you about the mass and size of a planet but says nothing about its atmosphere. Finally, astrometry tells you just about everything you want to know about a planet, but nothing about its atmosphere.

Interstellar Travel

But even when we do find the perfect candidate, it may still be a shot in the dark to know in advance whether they are truly habitable or not. The most difficult thing to determine will always be the exact atmospheric constituents, because due to the orbit and or brightness of the planet, spectroscopic studies may be impossible to carry out.

Planetary Moons

Perfectly habitable planets may be hiding-out in orbit around giant planets that are in the habitable zones of stars. We overlook them because we have no way to detect them. Their masses are completely masked by the masses of the giant planets that they orbit, but some astronomers think we may eventually find a few of these moons using the transit method. If the geometry is exactly right, you will see a dip in the star light from the giant planet followed quickly by a second smaller dip from the trailing moon. You may also see the dip from the moon followed by the bigger dip from the giant planet. Astronomers even have a name for these: Exomoons!

The star 1SWASP J140747.93-394542.6 is located 434 light years from Earth and is a very young, cool star only about 16 million years old. What is exciting about it is that the J1407 system experiences some very complex and unusual light eclipses. The current model that explains the details is that a small companion planet about 20 times the mass of Jupiter orbits this star at a distance of about 4 AU; nearly the orbit of Jupiter in our solar system. But this planet called J1407b has a massive Saturn-like ring system about 180 million kilometers in diameter. Also, this ring system has a gap about 61 million km (0.4 AU) from J1407b that astronomers consider to be indirect evidence of an exomoon with mass up to 0.8 Earth masses.

A second exomoon candidate has been spotted in the planetary system of the star WASP-12. The exoplanet WASP-12b was discovered in 2008 by the transit method, and has a mass of about 1.4 times Jupiter, but it orbits its star in a hellish orbit

Interstellar Travel

every day. The possible moon has a mass believed to be between one-half Earth and six-times Earth…but who would want to live there?

In time, after many more of the thousands of transit observations have been re-done by more powerful telescopes, we may uncover dozens if not hundreds of other planetary moons. Then we can really start to know how many habitable Earth-sized worlds exist among the nearest stars. Even now, Harvard astronomers are combing through the Kepler data to find more of these double-transit signatures in the existing data.

Oh, by the way, for you artists, please stop drawing these spectacular scenes that show a huge planet rising over the horizon. This could never happen physically, because the tidal gravity would have ripped its moon apart long ago! For example, our moon would be disrupted if it appeared a breath-taking, 20-times its full-moon size in our sky.

Space is not at all empty!

Space is not at all empty!

Contrary to what you might imagine, space is not empty. We know from nightly meteor displays that it is filled by small nuggets of matter. These rain down on the surface of Earth at a rate of 100 to 300 tons of cosmic dust every day. This dust is very small, microscopic in fact, but there is so much of it that it outweighs the occasional golf-ball or watermelon-sized meteor that also pelts our world every day.

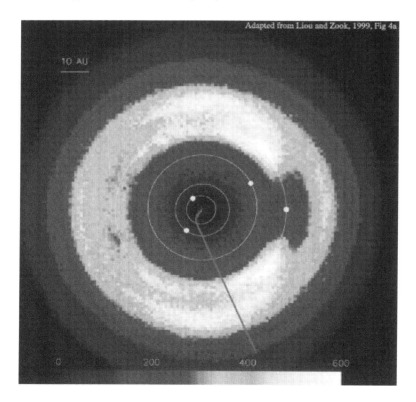

Interstellar Travel

The Student Dust Counter (SDC), engineered by students at the University of Colorado, is attached to the New Horizons spacecraft measuring dust on its way to Pluto. This image is a computer simulation of the possible dust distribution in our solar system based on the SDC data. The bar at the bottom shows the number of particles per AU^2. The green line shows the path of the New Horizons spacecraft. The SDC instrument can detect dust grains with a mass between on billionth and one trillionth of a gram. Even so, the figure below shows that over the course of the 8-year journey, the New Horizons spacecraft SDC detector, which is about 0.1 meters^2 was struck by one of these dust grains every week.

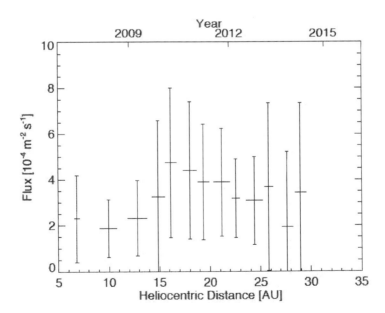

Space is not at all empty!

These micrometeors are created in the asteroid belt and in the tails of comets, and this dust fills most of the entire volume of the solar system. We can see the glow of this material in reflected sunlight called the Zodiacal Light. Earth plows through this debris and for the larger dust particles the size of rice grains, we get meteors in the night sky. But for the smaller dust grains we see nothing at all because the micrometeors barely make a flash as they penetrate the atmosphere. But they can be detected by radar. All of this interplanetary dust would amount to the same material in an asteroid about 30 kilometers across, but the solar system has had billions of years to generate this dust by asteroid collisions, comet evaporation, and some of it even comes from interstellar space!

From 2002 to 2006 the NASA Ulysses spacecraft measured interstellar dust grains as they impacted the spacecraft's sensitive dust detector. In the outer solar system, the interstellar impactors can be identified by their impact direction: they approach from a direction opposite to the rotation of the solar system. These dust grains have masses between 10^{-20} to 10^{-11} kg, with most near 10^{-16} kg. By studying dust clouds in interstellar space, astronomers have estimated the micron-sized dust grains have average masses near 10^{-16} kg.

Here we see a scanning electron microscope image of an interplanetary dust particle. It's very rough and seems to be aggregates of large numbers of sub-micrometer grains clustered in a random order.

Interstellar Travel

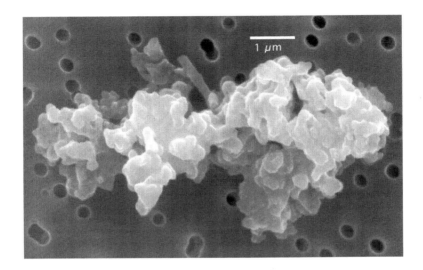

Farther out in space, the Voyager 1 and 2 spacecraft have studied dust impacts as far away as 60 AU from the sun. The spacecraft encountered micron-sized dust particles with masses of about 10 trillionths of a gram, with a density of about 2×10^{-8} particles per cubic meter. The most likely sources for these particles are comets. As the table below shows, both Voyager 1 and Voyager 2 (in parenthesis) noted a sharp fall-off in these dust particles beyond the orbit of Pluto between 30 and 50 AU.

Radial Distance	Impacts	Daily rate
6-15 AU	5 (7)	48 (46)
15-24	5 (19)	75 (111)
24-33	2 (17)	39 (99)
33-42	8 (0)	135
42-51	7 (0)	78
51-60	0 (0)	0

Space is not at all empty!

In addition to dust grains, there is also a thin, dilute plasma called the solar wind that travels at up to 400 km/sec leaving the sun. This is mostly made of individual protons and scattered atomic nuclei ejected from the solar surface in a constant stream, and rarely amounts to more than a few dozen particles per cubic centimeter. In its spiral path into the depths of the solar system, it brushes by the planets and compresses their magnetic fields. This is noticeable for Earth, Mars and Mercury, but Venus has no magnetic field. Instead, the wind makes direct contact with the atmosphere creating a comet-like shape as the Venusian atmosphere is dragged behind the wind. It takes almost a full year for the solar wind to reach the outer limits of the solar system where it finally slams up against a billion-kilometer shock front as it meets the interstellar medium itself.

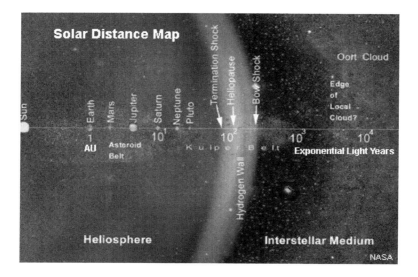

The edge of our solar system is called the heliopause, which is where the solar wind and the interstellar medium are in

Interstellar Travel

pressures balance. By the time we get to about 150 AU from the sun, we have left all traces of our solar wind behind and are now firmly within the grasp of the interstellar medium. Instead of a handful of atoms per cubic centimeter within the heliosphere, we encounter as little as 0.01 atoms per cubic centimeter. This low-density condition persists for many light years surrounding the sun.

For the next light year we enter the Oort Cloud – a vast halo of icy bodies kilometers across that were ejected from the solar system by their gravitational interaction with Jupiter when the solar system was formed. There are literally trillions of these bodies. Occasionally their nearly circular orbits are disturbed and they plunge into the inner solar system to become long-period comets. Comet Hale-Bopp may have been one of these.

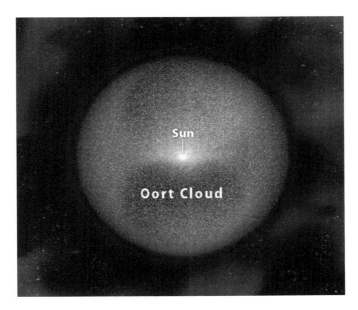

Space is not at all empty!

Although there are believed to be many of these, the volume of space out to several light years is so vast the average distances between comet bodies is probably billions of kilometers. They also move with speeds appropriate to their distance from the sun of less than 1 km/sec. Whether these comet nuclei have bumped together to generate their own dust clouds is unknown, but not expected. Instead, the space between them is probably dominated by the normal interstellar dust and gas. Beyond the Oort Cloud, astronomers have known for several decades that our solar system exists within a vast cavity in the interstellar medium created over 100,000 years ago when a nearby supernova exploded and evacuated this region of space. Normally, the average interstellar medium consists of about one

hydrogen atom per cubic centimeter, but in the so-called Local Bubble, the average density is only about 0.05 atoms/cc. This bubble is about 300 light years across, but it is not completely empty. It seems that there are filaments of gas clouds several light years across, and our solar system is near what is called the Local Fluff. This cloud has a slightly higher density of gas closer to 0.3 atoms/cc and in another 50,000 years our solar system will enter the outer edge of this cloud.

Volume of sphere 1 light years in radius:

$V = 4/3 \, \pi \, (9.5 \text{ trillion km})^3$
$= 3.6 \times 10^{39} \text{ km}^3$

Distance between comet nuclei:

$D3 = 3.6 \times 10^{39} / 1 \times 10^{12} = 3.6 \times 10^{27} \text{ km}^3$

$D = 1.5$ billion kilometers

$1 \text{ AU} = 150$ million km

$D = 10$ AU.

Interstellar Travel

We don't know if the average density of interstellar dust grains will change very much within the Local Bubble, but once we leave the Local Bubble some 300 light years from the Sun, we return to a more normal interstellar medium where the gas density is close to 0.5 atoms/cc, and about one out of every trillion particles in space is a dust grain.

These dust grains are produced from countless thousands of old red giant stars whose upper atmospheres were cool enough for silicate and carbon dust to condense like rain and be ejected into space by radiation pressure. There are also noticeable dark clouds that have accumulated over time, where the density of dust grains is thousands of times higher than in interstellar space. At these densities, and in clouds spanning a light year or more across, they completely extinguish the light from more distant stars and appear as black, starless blotches in space. The nearest one of these dark clouds is called the Taurus Dust Cloud and is located about 430 light years from Earth towards the constellation Taurus.

The following image shows the dense dust clouds near our solar system. The width of the figure is about 1600 light years. The right edge of the figure is in the direction of the Galactic Center and the constellation Sagittarius. The left edge is in the anti-center direction towards the outer Milky Way in the direction of the constellation Orion. The dashed line is in a direction nearly perpendicular to the disk of the Milky Way and has very little gas and dust. Astronomers think the supernova that created the Local Bubble actually blew out material along this path of least resistance forming a chimney!

Space is not at all empty!

If you are a celestial navigator leaving our solar system, there are definitely better directions to travel to avoid encountering interstellar dust grains. Do not go anywhere near dark dust clouds with their unknown hazards, and try to stay inside the Local Chimney if you can! So, what difference does dust and a few stray atoms make for interstellar travel? At the kinds of speeds we need to travel to make our journeys in a reasonable amount of time, they can make a huge difference. It actually comes down to just three things: 1) the kinetic energy of an individual particle, 2) the number of particles the ship will encounter at the speed it is traveling, and 3) the amount of damage or radiation one of these impacts will make.

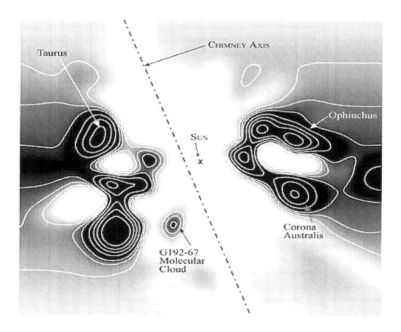

Interstellar Travel

The Ulysses spacecraft found dust grains with masses between 10^{-20} to 10^{-11} kg, with most near 10^{-16} kg. If our ship travels at 20% the speed of light, the kinetic energy of the dust grains is between 0.000018 Joules for the small grains, and 18,000 Joules for the large grains, with most kinetic energies being near 0.18 Joules. For a density of 2×10^{-8} dust grains per cubic meter, these dust grains arrive at a rate of 1 dust grain per second across a square meter of area.

> V=60,000,000 m/s
> M = 1x10^{-16} kg
>
> E=1/2(10^{-16})(6x10^7)^2
> = 0.18 joules
>
> Flux =(2x10^{-8} particles/m^3)x(6x10^7 m/s)
> = 1 dust grain/sec/m^2

What would one of the larger 18,000-Joule dust grain do, delivered in a one second impact to an aluminum hull? The heat capacity of aluminum is 0.9 Joules per gram. That means it takes 0.9 Joules to raise the temperature of one gram of aluminum by 1° Celsius. It's pretty obvious that this impact has the potential to raise a pinpoint-sized piece of the hull by 20,000° Celsius per gram! It would be like a hot knife through butter.

Another way to look at these impacts is by comparison with atomic-scale events. The 10^{-16} kg dust grain has a mass equal to 63 billion protons. The energy in joules per proton is 3×10^{-7} joules/proton, but when we convert this into atomic energy units where 1 electron-Volt equals 1.6×10^{-19} joules, we get nearly 2 billion volts (2 GeV as physicists would say). If these particles collided with the bulkhead material, they would be able to produce a pulse of gamma rays and secondary particles like muons and pi mesons that could penetrate deep into the surface of the spaceship, and also into the human habitation

Space is not at all empty!

environment! These larger dust grains are far from harmless, and at the speed of the spacecraft, you would get one of these pulses of radiation every second, for every square meter of area!

There are also the stray hydrogen atoms in the interstellar medium. In the Local Bubble, they amount to about 0.01 protons/cm^3. This is far higher than the density of dust grains. At a speed of 20% C, they carry a kinetic energy of about 3×10^{-12} joules, or in atomic units, about 180 MeV. At the spacecraft speed, they would pummel the bulkhead with F = (0.01/cm^3)*(6×10^9 cm/s) = 6×10^7 collisions per second for every square centimeter. You would essentially be traveling through a constant radiation field of 150 MeV particles, which are 10 times more energetic than ordinary cosmic rays! The secondary radiation entering the habitable areas would be significant and if unshielded, deadly over the long term. So, whether you think about interstellar dust grains or the interstellar hydrogen gas, you must deal with a significant source of radiation that must be shielded. The faster you go to minimize total exposure time, the more intense it becomes.

Although dust and gas are easy for astronomers to detect from Earth, larger objects the size of baseballs and planets roaming the interstellar void are impossible to see, and this presents a severe problem. If the spaceship encounters just one of these objects in its high-speed journey, it will be instantly fatal. They cannot be detected by telescopes in advance in the near-absolute darkness of space. Only some type of radar system would stand any chance of alerting the crew of an impending collision.

Interstellar Travel

NASA's main source of data about orbital debris in the size range of 1 to 30 cm is the Haystack radar. The Haystack radar, operated by MIT Lincoln Laboratory, has been collecting orbital debris data for NASA since 1990 under an agreement with the U.S. Department of Defense. Haystack statistically samples the debris population by "staring" at selected points in the sky and detecting debris that fly through its field-of-view. The data are used to characterize the debris population by size, altitude, and orbit. From these measurements, scientists have concluded there are over 500,000 debris fragments in orbit with sizes larger than one centimeter.

The 70-meter Goldstone antenna located near Barstow, California, when operated as a 400-kilowatt radar, is capable of detecting 2 millimeter debris at altitudes below 1,000 km. It has also been used to detect and study the moons of Jupiter and Saturn. Its maximum range is the orbit of Saturn for objects hundreds of kilometers across.

All of these installations are massive dishes, or equally large phased-arrays many tens of meters across, that generate megawatt pulses of radio-wave energy. If you are only interested in meter-sized objects, ESA has recently built and tested a small radar system for collision avoidance on the International Space Station and on other satellites. It can detect meter-sized objects at a distance of up to 500 km.

Meanwhile, the Arecibo radio telescope 300-meters in diameter located in Puerto Rico, is routinely used to study and image asteroids in the Asteroid Belt that are tens of kilometers across,

Space is not at all empty!

and it can detect objects tens of meters across out to the orbit of the moon.

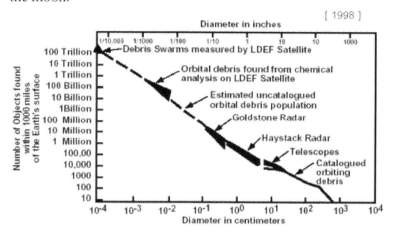

As these examples show, radar detection is a complicated balance between object size, detection range, and the transmitted power of the radar pulse. At a distance of a thousand kilometers, for a ship traveling at 20%C, you only have about 0.003 seconds before collision. In order to have an hour's notice, you have to detect the object at a distance of 1 billion kilometers (twice the distance to Jupiter!). That means with current-technology radar systems such as Goldstone with its 400 kilowatt radar, you will only be able to see objects larger than a few hundred kilometers. You will not see any of the smaller and equally-deadly interstellar debris until it is too late to take action!

Here is an example of a half-inch hole that was made in the Space Shuttle Endeavour's radiator panel by the impact of unknown space debris during STS-118 in 2007. The impact speed was typical of orbital debris of 10 km/sec. Space Shuttles and the ISS are constantly bombarded in Earth orbit by paint

Interstellar Travel

chips and other small debris only a few millimeters across, but traveling at 28,000 mph they are worse than common bullets.

Orbital debris is a hazard for catastrophic mission failure, and is one of the major worries of the International Space Station, whose large solar panels collect hundreds of collisions every year, degrading the performance of this vital electrical system over time. Although the near-Earth debris cloud is not directly relevant to interstellar voyages, it does point out how even rare but small impacts, especially at 20%C, can have devastating

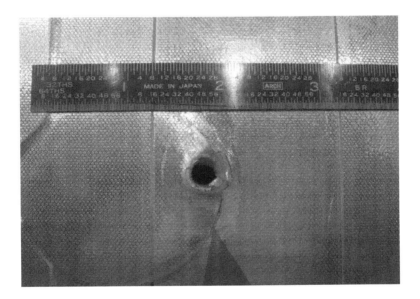

effects on the health and safety of such long-duration missions traveling through space, populated by unknown and undetectable hazards.

Space is not at all empty!

There is no technology on the drawing boards that helps interstellar spacecraft avoid the lethal effects of dust/asteroids impacts at very high speeds. The faster you go the more lethal are the interstellar dust grains, and the less time you have to react to the threat. There is no Earth-based technology that will let us detect meter-sized lethal bodies in space at distances of billions of kilometers allowing comfortable hours of warning time. Spacecraft will have to be constantly vigilant for these impacts, especially along the direction of travel where approach speed is highest, and unfortunately where the impact time is the shortest!

Interstellar Travel

Radiation

Radiation comes in two forms: particles and electromagnetic. Particle radiation includes protons, neutrons, electrons and the nuclei of many kinds of atoms such as helium and iron. Electromagnetic radiation includes all forms of light, which travel at 300,000 km/sec and include gamma rays, x-rays, ultraviolet, infrared and radio-forms.

Radiation in all its forms can be measured accurately. We can also, precisely determine at what levels it becomes a human hazard. Scientists measure radiation doses and dose equivalent in terms of units called Rads and Rems (Grays and Seiverts are used in the 'SI' System of meters-kilograms-joules).

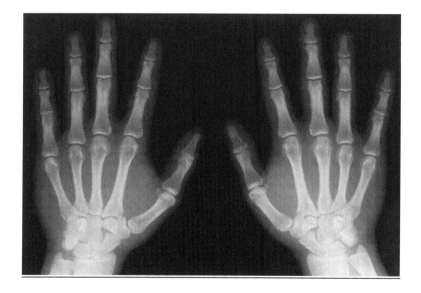

Interstellar Travel

Dose: This is a measure of the amount of total energy that is absorbed by matter over a period of time. This matter can be human tissue, or sensitive computer circuitry. The unit for dose is the Rad, which means 'Radiation Absorbed Dose'. One Rad is equal to 100 ergs (10^{-5} joules) of energy delivered to one gram of matter. The equivalent SI unit is the Gray (G). One Gray equals 100 Rads.

Dose Equivalent: This compares the amount of absorbed energy (Rads) to the amount of tissue damage it produces in a human. It is measured in units of the Rem, which means 'Roentgen Equivalent Man'. The equivalent SI unit is the Seivert (Sv). One Seivert equals 100 Rems.

Radiation dose is just the amount of energy delivered to a sample of matter. Equivalent dose, however, is much more complicated. This term has to do with the amount of damage that a given amount of energy does to a tissue sample or an electronic component. Each kind of radiation, for the same exposure level, produces a different amount of damage. Mathematically, this is represented by the equation:

Dose Equivalent (in Rem) = Dose (in Rads) x Q
or
Dose Equivalent (in Sieverts) = Dose (in Grays) x Q

Different forms of radiation produce different levels of tissue damage. EM radiation, such as x-rays and gamma-rays, produce 'one unit' of tissue damage, so for this kind of radiation Q = 1, and so 1 Rad = 1 Rem. This is also the case for beta radiation, which has the same Q value. For alpha particles, Q = 15-20, and

Radiation

for neutrons, $Q = 10$. That means that a dose of 1 Rad of radiation (which equals 100 ergs delivered to 1 gram of matter) produces a dosage of 10 Rem for $Q = 10$.

Dose Equivalent	Health Effect
50 - 100 Rem	No significant illness
100 - 200 Rem	Nausea, vomiting. 10% fatal in 30 days.
200 - 300 Rem	Vomiting. 35% fatal in 30 days.
300 - 400 Rem	Vomiting, diarrhea. 50% fatal in 30 days.
400 - 500 Rem	Hair loss, fever, hemorrhaging in 3wks.
500 - 600 Rem	Internal bleeding. 60% die in 30 days.
1,000 Rem	Intestinal damage. 100% lethal in 14 days.
5,000 Rem	Delirium, Coma: 100% fatal in 7 days.

Radiation comes from the sun and stars in the form of light, and also in the form of particles due to solar flares and other explosive events. Flares are produced when a star's surface magnetic fields become so twisted they re-configure themselves into simpler shapes. The resulting change liberates more energy than 1000s of hydrogen bombs. A careful study of stars like our sun or cooler reveal that they have magnetic activity cycles (sunspot cycles) similar to our own sun, but in some cases far more intense. Virtually all known stars that possess magnetic fields and activity have flares and other energetic events. Many stars cooler than our sun, called flare stars have 'solar flares' many thousands of times more intense than anything ever seen on our own Sun. Proxima Centauri, everyone's favorite star for a first interstellar voyage, is one of these dwarf flare stars. Solar or stellar flares are more than a curiosity and a nuisance. We have had many close-calls during the Space Age with intense

Interstellar Travel

solar flares that were unpredictable, but fortunately no one was in harm's way.

Radiation comes from the rest of the universe in the form of cosmic rays created by supernova explosions and by certain kinds of galaxies in deep space. Radiation also comes from radioactive minerals such as feldspar (uranium and radon), which is common on the surface of granite-rich planets.

There are also two distinct environments in which travelers will be exposed to these radiations: in space and on a planetary surface.

Planetary Surfaces - On the surface of a planet with an atmosphere, the atmosphere itself provides an enormous natural shield to particles radiation, though even on the surface of Earth we still get a measurable amount of cosmic rays, especially if you live at higher elevations.

The amount for shielding is just the shear mass of atmosphere above your head. For Earth, we have about 240 gm/cm^2. For a planet like Mars with 1/10 the atmosphere of Earth, there is hardly any shielding at all. It's about 12 gm/cm^2, or about the same as the walls of the International Space Station. Mars is popularly considered colonizable even though no colonist could ever stroll its surface without the aid of a spacesuit equipped with radiation shielding. On Earth, the cosmic ray exposure is about 26 milliRem or 7% of your annual total at sea-level, and about 50 milliRem or 14% at the elevation of Denver. Typically, our annual radiation dose is about 370 milliRem (or 3.7

Radiation

milliSieverts). On Mars, the average exposure is about 20 milliRem per day!

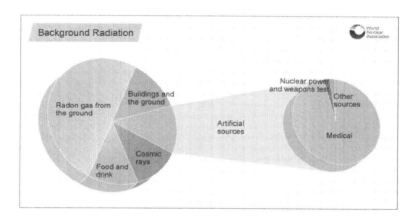

Also on the surface of Earth, we have granite-rich rocks in the crust that emit radon gas providing a background dose that we can only minimize by living in areas where there is little of this granite-rich material. On Earth, 50% of your annual radiation exposure comes from radon gas generated by clays and building materials! For example, if you decided to live outdoors in a tent, your exposure would be 0%, but if you lived in an unventilated basement recreation room, it would be 50% of your total natural background exposure. Radon exposure is responsible for up to 14% of all lung cancer cases.

Electromagnetic forms of radiation such as x-rays and gamma-rays are shielded from the planetary surface also by virtue of their being a thick atmosphere. With the exception of ultraviolet, and some forms of infrared and radio radiation, a

Interstellar Travel

typical atmosphere is thick enough to support a humans and a biosphere, however the exception is ultraviolet light. The most lethal forms can only be blocked by the presence of an ozone layer produced by free oxygen. Without an ozone layer, UV-B radiation reaches the surface and is eventually lethal to exposed, unshielded life. Life can still exist in the water, which provides enough shielding for ultraviolet only a few meters below the surface, but land-based life must operate in the night-time and has to be protected during the harsh daytime. Organic optical sensors (eyes or photoreceptors) will quickly form cataracts if unprotected for long periods of time. UV-B is also a known mutagen so prolonged surface exposure leads to the eventual accumulation of harmful mutations unless they can be repaired quickly. Meanwhile, there are no known problems with infrared and radio emissions that reach the ground.

In Space - Here we have to deal with cosmic rays during the voyage, and both cosmic rays and high-energy particles generated in stellar flares once we reach our target planetary system.

One of the closest sources of cosmic radiation other than our sun is the Crab Nebula; the remains of a massive star that exploded in 1054 AD at a distance of 6,500 light years.
Cosmic rays are far more energetic than other kinds of particle radiations emitted by stars, because they have been generated by the energy of entire exploding stars and then accelerated for millions of years to enormous energies as they circulate around the entire Milky Way. Cosmic rays from a supernova are like the energy emitted by a camera's flash, but the highest energy

Radiation

cosmic rays are created by a process similar to what scientists use in 'atom smashers' like the Large Hadron Collider.

In deep space far from a star, you will experience the full-bore intensity of cosmic rays. In the solar vicinity out to many light years, cosmic ray intensity is what we call 'isotropic' because it is about the same in all directions. There are no favored directions you can travel that minimize its intensity because cosmic radiation will penetrate your living quarters from all directions. This map shows where cosmic rays come from across the entire sky.

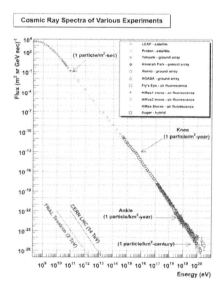

Cosmic rays also come in a variety of different energies. When physicists discuss them, they find it convenient to describe their 'energy spectrum', which gives the flux of cosmic rays at various energies. This instantly tells a physicist how the energy is spread out, and how often you should expect to detect a cosmic ray particles of a particular energy. This figure, for example, shows that at an energy of '10^{11}' electron volts (eV), you should expect to detect about one particle every second in an area about 1 meter². Since 1 eV equals 1.6×10^{-19} joules, the cosmic ray carries about 1.6×10^{-8}

Interstellar Travel

joules of energy per hit. This doesn't sound like much, but in atomic terms it is HUGE. It equals 100 GeV, and these kinds of collisions at laboratories such as CERN and Fermilab lead to rain storms of hundreds of secondary particles.

So, instead of having to block or shield yourself from just one particle, you have to shield against hundreds of secondary radiation particles after they hit the wall of your ship...or your spacesuit!

Shielding

The intensity of radiation can be reduced in many different ways depending on the type of radiation and its energy. For example, ordinary visible light can be shielded by any opaque substance, but x-rays require a denser medium. Similarly, various forms of matter radiation can be shielded by increasing the number of particles they have to scatter off of as they pass through a medium. Each time a particles of radiation scatters off a

Radiation

shielding particle it loses some of its energy, and the farther it penetrates, the more energy loss occurs.

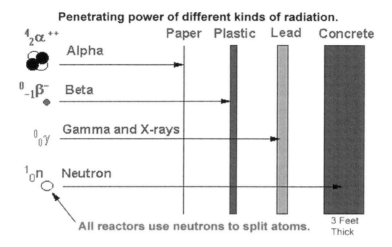

Penetrating power of different kinds of radiation.

As a general rule, the more mass you can put between you and the source of the radiation, the better off you are. But there is a problem. For very energetic particles of radiation, when they interact with the shielding material, they can emit secondary radiation particles that, themselves, can be a problem. For the highest-energy cosmic rays, the more shielding you use, the more radiation you generate inside the shielding, so now you have to find an optimum shielding thickness that reduces the dosage from the secondary shielding radiation too! This is why designing the proper shielding for a spacecraft can be a tricky proposition, and you really have to know the energy spectrum of the particles you are trying to shield against to get things right.

Biophysicists who measure and worry about the radiation health effects to humans work with a selection of materials used in

Interstellar Travel

designing spacecraft bulkheads, spacesuits and habitation modules to come up with the best design for radiation shielding given the expected radiation levels that the crew is exposed to during a particular period of time. The basic idea is to first decide on what the maximum expected radiation dose can be over the exposure time of the crew, which does not exceed established healthy levels. Once you know the target level on the human-side of the shield, you then predict what the expected environmental-side looks like in terms of types of radiation and their energies. These two numbers tell you by what factor you have to reduce the radiation flux with the shielding material. The last step is to find an appropriate shielding material and thickness that meets the size and mass constraints set by the engineers. Here are some examples:

Spacesuits and EVAs – The amount of shielding is about 0.3 gm/cm^2 and this is enough to protect an astronaut from most forms of cosmic radiation of moderate energy for up to several hours. However, a major solar flare that occurred on August 2 1972 would have exposed a lunar astronaut to over 500 Rems of radiation in the spacesuit, which would have been life-threatening without prompt medical attention – but not necessarily fatal.

International Space Station – The shielding from the average bulkhead is about 5 grams/cm^2. The average radiation dose for ISS astronauts working in space for 6 months is 0.05 to 2 Seiverts (5 to 20 Rems). However, on January 20, 2005 a powerful solar flare called NOAA 720 bathed the ISS in high-energy 100 MeV proton radiation for several days. An astronaut on EVA would have gotten radiation poisoned, but the ISS crew

Radiation

inside received only about 1 Rem of exposure thanks to the bulkhead. This is the same exposure they would have received living on Earth in two years, but on the ISS they received the same dose in a matter of a few days.

Trip to Mars – The transit vehicle will provide about 10 to 15 gm/cm² of shielding. The Mars rover Curiosity has allowed us to finally calculate an average dose over the 180-day journey. It is approximately 300 mSv, the equivalent of 24 CAT scans. In just getting to Mars, an explorer would be exposed to more than 15 times an annual radiation limit for a worker in a nuclear power plant. Once on the Martian surface, cosmic radiation coming from the far side of the planet is blocked. But on the day-side of Mars, protection from strong solar particles varies considerably as the thin Martian atmosphere is bombarded by the solar wind. Although Earth's atmosphere is equal to a shielding of 1000 gm/cm², the thin Martian atmosphere provides only about 16 gm/cm². That means on the surface, the unshielded radiation dose rate can vary from 10 to 20 Rems/year, which is over 30 to 60 times higher than on Earth.

This brings up another issue with radiation: the rate of exposure. 1 Rem of radiation delivered over a year or more over your entire body surface area and volume can be completely harmless, but the same radiation delivered in a few seconds can be a major hazard. Radiation damages cells, but cells can repair themselves given enough time. The problem occurs when the radiation rate is so high a cell does not have time to complete one repair before the next damaging event happens and the repair system is overloaded. This happens at about five weeks. This sounds like a long time per cell, but at a high enough

Interstellar Travel

radiation level, the same cell can be struck by a radiation event multiple times within this five-week time, causing the repair mechanisms to fail resulting in cell death.

Cells are not the stable entities you think they are. Every hour over 10,000 measurable DNA changes can happen in a single cell every hour. Thankfully, we have many protective repair mechanisms that constantly look for damage and then take the necessary molecular steps to return the organism to a functioning state. Under a radiation exposure of 1 centi-Grey (1 Rad) about 100 DNA changes occur. If we convert this in to an annual repair rate, we get about 90 million DNA changes per cell per year from ordinary biological activity, and for the average ground-level dose of 0.24 cGy/yr we get about 24 changes. So under normal conditions, the radiation background is utterly irrelevant.

Overall, the response of an organism to an acute radiation dose or an increase in dose rate is generally described by a dose-response function that changes from low-dose benefit into high-dose harm, at a specific threshold called the NOAEL, which means 'no observed adverse effects level'. For a short-term radiation dose, ~ 50 Rem (0.5 Gy), and for a lifetime dose rate, ~ 0.7 Gy per year.

We can adjust our shielding to get to Mars and back in 500-days so that the total dose from cosmic rays is well below NOAEL, but for interstellar travel we have a problem: We don't really know what the cosmic ray rate looks like outside our solar system! We can guess, but if we are wrong by a factor of two, we will end up with travelers in pretty bad shape when they get to

Radiation

their destination. Also, the more time you spend in space, the worse your radiation exposure will be. The only thing we know for sure is that we need to keep the dosage to under 0.7 Gy/yr (70 Rads/year).

We do not expect any short-term events like solar flares, so that makes the calculations very simple. All we have to do is to figure out the average cosmic ray background over a multi-century trip, in a region of space for which we do not have any direct cosmic ray measurements yet.

When we reach our destination, we have other issues too. The biggest one is that the star may produce intense flares from time to time. Luckily we can study these stars from Earth to see which ones are 'active' and then either anticipate the radiation shielding we will need when we get there, or even avoid that destination in the first place.

Solar flares often roil the surface of our sun due to complex magnetic fields, but are harmlessly dissipated and blocked by Earth's atmosphere. This is not the case for astronauts operating in space. This image shows sun erupting with an X1.7-class solar flare on May 12, 2013. It was taken by NASA's Solar Dynamics Observatory. Other stars appears to be far more active that our sun as sources of flaring activity.

As we discussed earlier, the ship will travel at high-speed through the hydrogen atoms (protons) of the interstellar medium, and this leads to a radiation exposure of 180 MeV per collision, and there will be 6×10^7 of these across every square centimeter, and every second, for typical interstellar gas densities

Interstellar Travel

in the Local Bubble of 0.01 atoms/cm³. This will actually be a higher exposure than ordinary cosmic rays. To make matters worse, most of the radiation will appear to come from the forward-edge of the spacecraft because of the high speed.

> $D = 0.53 \times N \times H \times d \times S/M$
> where N is the number of protons/cm²/sec striking the ship, H is the stopping power of 1 GeV protons in human tissue, d is the average thickness of a human torso, S is the average cross-section of a human body, and M is the approximate mass of a human.
> For:
> $N = 6 \times 10^7$ protons/cm²/sec, H=2 MeV/cm, d=30 cm, S=7000 cm² and M=70000 gm,
> You get:
> $D = 2 \times 10^8$ Rads/year, and with Q=10 for protons you get 2×10^9 Rems/year.

It is possible to calculate what the radiation dosage would be for an unshielded astronaut outside the spacecraft. The answer you get for a proton radiation is 2 billion Rems/year. This is a difficult number to appreciate but in terms of your dose per second it works out to about 60 Rems/second. This is about the same lethality as sitting in the middle of a nuclear reactor! In fact, the radiation you absorb will literally cook you alive at these energy rates! No amount of physical shielding will reduce this to levels near 1 Rem/year that can be survived without undue added cancer risk.

But perhaps there is something else we can try. Suppose we intercept the particles before they even reach the ship? A large mass positioned ahead of the ship like a shield would take the brunt of the impacts and harmlessly absorb or deflect them before they reached the ship. This adds mass to the ship, but

Radiation

perhaps it is worth the added weight to have a crew that survives to its destination!

Another way that would work for electrically-charged particles is magnetic shielding. This is actually being studied by NASA. Let's have a look at this elegant and very science-fiction-like solution and see where it takes us!

Magnetic fields as shielding

Astronauts can work in the International Space Station because not only is there still some dilute atmosphere to help with the shielding as well as the bulkhead of the space station itself, but because Earth's magnetic field deflects the highest-energy particles away from the inner Low Earth Orbit regions. Nevertheless, this magnetic shielding is not perfect and breaks down for the most energetic cosmic rays, which still make it to the ground. There are also 'air shower' events caused by the highest-energy cosmic rays that produce a steady rain of radiation at the surface.

A target planet with a strong magnetic field will be very helpful, and in fact recent studies of the evolution of Mars and Earth suggest that a strong magnetic field is essential for hanging on to an atmosphere against the steady wash of the solar wind from the companion star. Instead of solar wind particles helping to evaporate a planetary atmosphere, they are deflected away from the atmosphere by the planet's magnetic field. Mars does not have a magnetic field and this probably helped the planet lose most of its atmosphere over billions of years. It is likely that if we find a planet with a breathable atmosphere as a target for our

Interstellar Travel

travels, it will already have a magnetic field as a 'free-be' and so we will likely have this shielding available to us on the surface.

How about using magnetic fields to shield the ship from cosmic rays, interstellar gas and dust grains? They won't work for neutral particles like neutrons, but the good news is that dust grains are probably charged, and of course so are the interstellar hydrogen atoms (actually just bare protons) and cosmic rays. In principle all of these objects can be steered by magnetic fields, and this prospect has led to many interesting spacecraft designs.

One famous science fiction solution to the hazards of high-speed interstellar travel is the Broussard navigational deflector used on the starship "Enterprise" of *Star Trek* fame. It works both on sub-light impulse drive near 20%C, and in hyperspace!

Magnetic fields have been investigated many times as possible shields, but they always run up against the same problem. The intense field needed to do the shielding is lethal to humans!

To understand magnetic shielding is a bit trickier than the simple ballistics of dust grains and cosmic rays. A magnetic field with a strength of B, will create a force on a charged particle with charge Q and speed V according to the Lorentz Force Law

Radiation

$$F = Q V \times B$$

The direction of this force, F, will be perpendicular to both the direction of the magnetic field, B, and the velocity, V, of the charged particle, Q. For example, if a charged particle travels along your line-of-sight, and the magnetic field is oriented north-to-south, the force will be from east-to-west.

Another thing that happens is that the particle feels this as a centripetal force that tries to get it to spiral around the magnetic lines of force. Centripetal forces are calculated by another formula

$$F = MV^2/R$$

where V is the speed of the charged particle and R is what is called the gyroradius. When we put these two formulae together in balance and simplify with a little algebra we get

$$B = MV/QR$$

If we want a proton with a charge of $Q=1.6 \times 10^{-19}$ Coulombs to be deflected by $R=1$ meter traveling at $V=60,000,000$ meters/sec (20%C) with an energy of $E=150$ MeV where 1 MeV = 1.6×10^{-13} Joules, we get

$B = 2 (150 \times 1.6 \times 10^{-13})/(1.6 \times 10^{-19} \times 6 \times 10^7 \times 1)$
$B = 5$ Teslas.

But the kinetic energy of the particle is $E=1/2 MV^2$, so in terms of the particle energy we get

$$B = 2E/QVR$$

What this means is that if we want to shield ourselves from a 150 MeV over a thickness

Interstellar Travel

of a generous 1 meter bulkhead, we need a magnetic field of 5 Teslas. This doesn't sound like much, but when you consider that this field has to be maintained across the entire surface area of the spacecraft, this presents a severe problem. The total energy required is just $E = 8\pi B^2 \times$ Volume, so that a 5-Tesla field in a volume the size of a large building (50-meters x 50-meters x 100-meters) requires 1.6×10^{15} joules. This is as much energy as 0.4 megatons of TNT. Fields this strong, though occupying only a few cubic meters, are common in hospital MRI machines. The strongest continuous magnetic fields currently available are about 45 Teslas, and require about 45 megawatts of power to sustain, in a volume less than a cubic meter.

There have been many studies since the 1960's that have attempted to design spacecraft shielding using superconducting magnetic fields. The currents required to shield even a few hundred cubic meters of space are in excess of one million amperes for deflecting 100 MeV protons with field strengths between 5 and 10 Teslas. Because the coils are superconducting, you only have to energize them once for the trip, requiring 55 hours of charging using a 10 kilowatt power source. But the stored energy, if released because the superconducting system failed, would cause catastrophic melting of the spacecraft!

Mutations

Mutations in our genetic code are inevitable because even during the act of replication the billions of base-pairs in our genome can result in occasional misspellings of the A ,G ,T and C coding for specific proteins. These lead to the production of different proteins than required in a particular gene. Usually these errors can be repaired by processes in the cellular nucleus and are rendered harmless. But if this repair process is overwhelmed because you are in a high-radiation environment, mutations can persist. When these happen in ordinary somatic (body) cells, this can either lead to a dead cell, or one behaving badly like a cancer cell. When the mutations appear in a sperm or egg cell, they can be passed on to the next generation. This happens in cases like sickle cell anemia, where a single base-pair error leads to a protein that causes a malformed red blood cell.

Because women have two Y chromosomes and men have an X and a Y, genetic errors in the Y chromosome can often be repaired because the corresponding healthy Y chromosome has a good copy of the particular gene. For men, this natural redundancy is not available, and this is why men have more sex-linked diseases that shorten their lifespans than women do. In some sense, women are a better flight crew than men from the standpoint of genetic health. But there are other ways to insure that interstellar voyages and colonization leads to a healthy stock of humans that can survive despite the accumulation of genetic diseases and mutations.

A recent statistical model created by Cameron Smith at Portland State University in Oregon shows that the best population size

Interstellar Travel

for a healthy colony is about 20,000 to 40,000 people! This is a far cry from the few-hundred often imagined. In fact, there is now considerable archeological evidence that human 'bottle necks' in evolution occurred when the breading population fell below a few thousand.

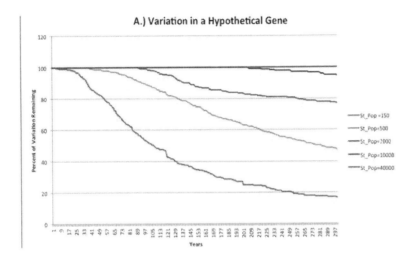

This number would maintain good health over five generations (100 years) despite (a) increased inbreeding resulting from a relatively small human population, (b) depressed genetic diversity due to the founder effect, (c) demographic change through time and (d) expectation of at least one severe population catastrophe over the five-generation voyage. Sending frozen sperm and eggs on the voyage with a limited number of human "tenders" is also an option, Smith said, though he didn't consider it seriously.

Mutations

A crew of 150 people is not nearly high enough to maintain genetic variation. Over many generations, inbreeding leads to the loss of more than 80 percent of the original diversity found within the hypothetical gene. A population of 500 people would not be sufficient either, and this scenario loses 45% of its original diversity. A starting population of 40,000 people maintains 100 percent of its variation, while the 10,000-person scenario stays relatively stable too.

When NASA's Curiosity rover was en route to Mars on a 253-day voyage through interplanetary space, its radiation sensors directly measured how 'hot' interplanetary space was and whether such a journey would be a hazard for future travelers.

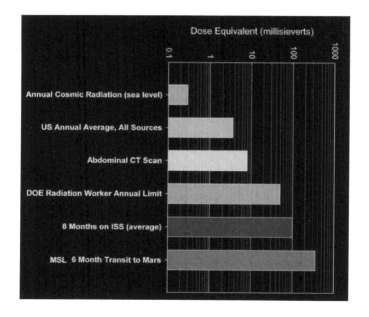

Interstellar Travel

This graph compares the radiation dose equivalent for several types of experiences, including a calculation for a trip from Earth to Mars. The data show that during a typical 6 month cruise to Mars the astronaut crews would be exposed to nearly four times the typical 6 month exposure of astronauts aboard the ISS. The spacecraft, though not designed to shield humans, was exposed to an average of 1.8 milliSieverts (180 milliRem) per day during the trip to Mars due mostly to galactic cosmic rays. The round trip dose of 9 Seiverts (90 Rads) doesn't even include the astronaut's surface stay on Mars, which is about 50% less than in space, thanks to what there is of the thin Martian atmosphere. It is hoped that added shielding can bring these dose numbers way down, while at the same time not adding huge costs and weight to the mission.

Radiation does more than 'merely' cause genetic damage. It can also cause an accumulation of brain injuries that over time affect cognition and memory. Researchers used to think that slow-growing brain cells were less vulnerable than fast-reproducing cells to radiation damage, but this now seems to be no longer true.

When high-energy, highly-charged particles enter brain tissue, they cause a narrow channel of cellular damage as they pass through the skull and exit. They also produce secondary particles as they collide with the nuclei of atoms in these brain cells. These secondary particles can take their own paths through the brain and damage other cells over time. According to a recent calculation by Marcelo Vazquez, a researcher at Brookhaven National Laboratory, large fractions of the cells in astronaut's brains would be hit at least once by these energetic

Mutations

particles on a 3-year trip to Mars. Astronauts with severely impaired brain cells could experience certain kinds of memory impairment and also become depressed. They might even develop conditions resembling Alzheimer's disease.

The relationship between effective population size and fixation index. Small populations become inbred more rapidly than large populations, often leading to inbreeding depression. As the effective population size (N) becomes smaller, it becomes more likely that individuals in a population will mate with relatives. As a result, small populations experience an increased degree of inbreeding (F increases) with lower levels of genotypic diversity. Inbreeding depression is of great concern for wild animals on the brink of extinction. It will also be a concern for interstellar travelers.

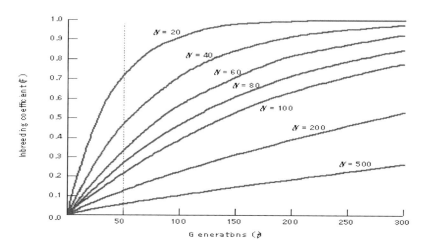

A single base pair mutation in the human genome results in blond hair and blue eyes because of a melanin deficiency, but

Interstellar Travel

sometimes the process goes awry. The resulting blue eyes in people with plenty of melanin is especially dramatic, but entirely harmless. Other single base-pair errors are far more problematical such as sickle-cell anemia.

The main problem we have to face with human voyages to the stars is that interstellar travel will produce mutations in the human genome that will have to be ruthlessly purged in order not to diminish the future viability of the human gene line, thereby defeating the purpose of the entire trip. There is no religion we will take with us, or moral code we recognize today, that will allow us to terminate the number of pregnancies where the genetic mutation load is mathematically deemed to be too high... no more said.

Diseases

A disease is a particular abnormal condition, or a disorder of structure or function, that affects part or all of an organism. Disease is often construed as a medical condition associated with specific symptoms and signs. There are two basic forms of diseases: genetic and environmental.

Genetic diseases – stem from mutations in genome that alter human-normal cellular functions. Over 10,000 human diseases are known to be 'monogenic'. Pure genetic diseases are caused by a single error in a single gene in the human DNA. The nature of disease depends on the functions performed by the modified gene. The estimated 23,000 genes that make up a human being are arranged along tightly bundled strands of a chemical substance called deoxyribonucleic acid, or DNA. The DNA strands are tightly packed into structures called chromosomes. Over 1,000 known disorders are caused by chromosome abnormalities.

In 1957, Victor McKusick was appointed director of the new Moore Clinic for Chronic Diseases at Johns Hopkins University In 1966, he published his first catalog of all known genes and genetic disorders, *Mendelian Inheritance in Man* (MIM). More than 6,000 single gene disorders are currently known, meaning that mutations in somewhere around 24% of the approximately 25,000 human genes found so far can cause genetic disease. Currently 39% of all the disease genes found have been discovered on chromosomes 1, 6 and X, which are called the human 'disease rich' chromosomes.

Interstellar Travel

Although many diseases are the result of a single gene mutation, a number of important diseases result from several genes malfunctioning such as diabetes, asthma, cancer and mental illness.

Environmental diseases – stem from exposure to an external toxin or other invading organism found in the environment.

Bacterial and viral organisms can invade the human body and usually trigger several kinds of immune responses, beginning with fevers and inflammation, which is a primitive way for dealing with invaders. White blood cells usually lead the charge and are effective for many kinds of simple infections once the invading cells are sensed. A second immune system involving lymphocytes (T-cells and B-cells) and antibodies can identify invading cells by receptors on the invading cellular surfaces. Once the antibody attaches to the cell, it can be sensed by lymphocytes in the blood that then attack the cell and destroy it. If you have been exposed to this invader before, you already have antibodies for it and this gives you a degree of 'immunity' from the invading disease. If not, you may still have some randomly-present antibodies for it at a very low level and so your immune system has something to start with as it marshals its response more effectively. If you have no antibodies, you can get an inoculation before you contact the disease that presents dead invader cells to your immune system, which then goes ahead and builds up a supply of antibodies to the invader. Modern flu vaccines change every year as the invading flu virus mutates and its surface proteins also change. The vaccine lets the immune system catch up to the flu and develop antibodies for it.

Diseases

Streptococcus bacteria were found on the Surveyor spacecraft and brought back to Earth where they were revived after three years in a vacuum.

Although bacteria are organisms that can live and reproduce by themselves because they carry DNA, viruses are far-smaller and usually carry only RNA. They require a host to invade and hijack its reproductive system. The smallest virus has only about 500 base-pairs in its RNA 'genome'. Millions of bacteria normally live on the skin, in the intestines, and on the genitalia. The vast majority of bacteria do not cause disease, and many bacteria are actually helpful and even necessary for good health. These bacteria are sometimes referred to as "good bacteria" or "healthy bacteria." Harmful bacteria that cause bacterial infections and disease are called pathogenic bacteria.

Common pathogenic bacteria and the types of bacterial diseases they cause include:

Escherichia coli and Salmonella cause food poisoning.
Helicobacter pylori cause gastritis and ulcers.
Neisseria gonorrhoeae causes gonorrhea.

Interstellar Travel

Neisseria meningitidis causes meningitis.
Staphylococcus aureus causes a variety of infections in the body, including boils, cellulitis, abscesses, wound infections, toxic shock syndrome, pneumonia, and food poisoning.
Streptococcal bacteria cause a variety of infections in the body, including pneumonia, meningitis, ear infections, and strep throat.

There are over 30 bacterial diseases that are common in humans. Most can be treated with bacteria-killing antibiotics like penicillin, streptomycin and vancomycin, however, bacteria like all organisms evolve in response to their environment. This means that improper use of antibiotics can gradually lead to populations of bacteria that are immune to the antibiotic. It doesn't take long for this immunity to be passed-on either, even for terrestrial organisms. Bacteria are large enough that their entry into the body can often be prevented through simple acts like washing hands, using a face mask, or consuming food that has been heated so that bacteria are killed above 150 F. They are often transmitted through direct physical contact or exhaled droplets during respiration or sneezing.

Although bacteria can usually be blocked by the cell membrane, viruses commonly invade cells and hijack their reproductive system to replicate. They are the most primitive and numerous form of life/nonlife on the planet, and a constant source of concern. Although bacterial diseases are rarely air-born, many viral infections such as the common flu and the 'cold' are airborne and travel rapidly through a population. Symptoms may not appear until long after the host has incubated the

Diseases

disease, the virus has replicated itself inside the host, and has been re-transmitted back into the population through respiration.

Although bacterial diseases are fought by chemically destroying their cellular membrane or interfering with their reproductive cycle, viruses often have to be fought by preventing them from invading the host cell in the first place.

To find an antiviral drug, you have to first isolate the virus causing the disease. This can be a challenge. Once the virus is identified, a process of identifying its unique surface receptors or proteins required for reproduction are isolated and a new protein 'antibody' could be created chemically to attach to the virus so that the immune system can identify it and destroy it. Sometimes it is just easier to create a vaccine by letting a large number of these viruses reproduce, then kill them and inject them into the patient, whose own immune system then senses these deactivated cells and creates antibodies for them.

In any event, none of the development of vaccines or other antimicrobial therapies can begin until someone becomes infected and the disease pathology can be identified. For colonists visiting a planet with a biosphere, it cannot be known in advance which of the literally millions of types of bacteria and viruses native to that planet are going to be pathogens, or merely give the visitor a bad cold.

Does the fact that our DNA may be different from the alien biosphere's DNA help? Our current biospheric DNA is the result of literally billions of years of evolution unique to the conditions on this world. Would an alien biosphere have similar

Interstellar Travel

families of bacteria and viruses as Earth? The remains an interesting and open question, but the survival of human visitors to these distant biospheres may depend on the answer. Another issue is the reproduction time for a bacterium or a virus.

Bacterium	Time (minutes)
Escherichia coli	17
Bacillus megaterium	25
Streptococcus lactis	48
Staphylococcus aureus	30
Lactobacillus acidophilus	87
Mycobacterium tuberculosis	932
Treponema pallidum	1,980
Mycobacterium Leprae	43,200

Viruses on the other hand can lay dormant in the host for many years before they are triggered into action. For example the common herpes virus becomes active under stresses. There is even some evidence that there are viruses that lay dormant for decades. The following table lists only a few of the many human-targeted viruses that have been recognized as 'slow viruses'. There are many more that affect other mammalian species.

Virus	Disease	Latency
Rabies	Rabies	3 to 12 weeks
HSV-1 and 2	Herpes Simplex	Months to years
HIV	AIDS	5 to 10 years
Measles	Sclerosing Panencephalitis	10 years
Rubeola	Rubella Panencephalitis	10 to 20 years
HTLV-1	T-cell Leukemia / Lymphoma	10-30 years
HTLV-2	Atypical Hairy Cell Leukemia	10-30 years
???	Multiple Sclerosis	Decades
JC Virus	Multifocal Leukoencelopathy	Years to life

Diseases

There are also diseases produced by biological particles even smaller than viruses called prions. These particles do not even have RNA or DNA and they work by simply by causing other proteins to miss-fold themselves. A protein's shape is an important 'lock and key' element to how molecules work, and when shapes are distorted, the function of the protein dramatically changes. Examples of prion-mediated diseases are Transmissible spongiform encephalopathies (TSEs), including Kuru and Creutzfeldt–Jakob disease of humans, scrapie of sheep, and bovine spongiform encephalopathy (BSE) of cattle, were previously classified as slow virus diseases as well.

Some diseases are caused by invading microorganisms and microscopic parasites like protozoa and nematodes among many others. Parasites may invade a human in search of food (tapeworms) or as part of their reproductive cycle (Guinea Worm). Protozoa are single-celled organisms that can reproduce rapidly and interfere with organ function in the victim. Malaria is spread by mosquitoes that are infected by the parasites *Plasmodium falciparum*, *P. ovale*, *P. vivax* and *P. malariae*. In most cases you avoid these diseases by avoiding the ecosystem where they breed, or by creating physical barriers to them (breathing masks, water treatment, or separating drinking water sources from waste disposal).

What happens to the human immune system when it is placed in a sterilized environment? The growing incidence of children with allergies is a recent example of the over-use of antibiotics in the bathroom, and children no longer playing outdoors in the numbers that they used to.

Interstellar Travel

The results of two NASA investigations suggest that spaceflight may temporarily alter the immune system of crew members flying long-duration missions aboard the International Space Station. Researchers examined the blood plasma of 28 ISS crew members before, during and after their missions. Although the distribution of immune cells in the blood of crew members is relatively unchanged during flight, some cell functions are significantly lower than normal, or depressed, while some cell activity is heightened. In a sense, the immune systems of crew members become confused.

The immune system is likely being altered by many factors associated with the overall spaceflight environment. Radiation, microbes, stress, microgravity, altered sleep cycles and isolation could all contribute to this immune system decline. If this situation persists for longer deep space missions, it could possibly increase risk of infection, hypersensitivity, or autoimmune issues for exploration astronauts. There doesn't seem to be any time during spaceflight where researchers see stabilization of the immune system.

Meanwhile, many pathogens have a great time in space, growing stronger and increasing their resistance to antimicrobials. Both herpes and staph have been shown to thrive in the gravity-free, hyper-sterile environment of a space vessel.

The NASA research suggests that part of the problem seems to be that certain critical genes that activate the immune system are not turning on under microgravity conditions. Normally, a signaling pathway called PKA responds by turning on 99 genes that then go on to activate the immune system's T-cells to fight

Diseases

invading viruses and bacteria. But under microgravity conditions, 91 of these 99 genes remain turned off.

Another problem is that all people normally carry latent viruses, which a healthy, functioning immune system usually keeps in check. But in space, some of these viruses, such as herpes and Epstein-Barr, can reactivate and cause disease. Even worse, some bacteria with antibiotic resistance seem to thrive in the space environment. The extra radiation in space also weakens the immune system, and it is possible microgravity and radiation cause more damage to the immune system together than they do separately.

At the present time, there is no reliable evidence that suggests an astronaut with a suppressed immune system would be able to complete a three-year space mission - the estimated time for a round-trip to Mars. To study this problem, identical twins Scott and Mark Kelly will be monitored for a full year in 2015. Both are astronauts, and genetically identical, but Mark will remain on Earth as a reference to see how his brother's immune system on the International Space Station changes.

The figure on the next page shows the complexity of the immune system with an overview of the T-dependent humoral immune response. The diagram shows many of the key cells involved in the body's reaction to antigens. Also shown are the principle surface receptors. Astronaut studies will help reveal exactly how the human immune system 'shuts down' and whether there are any possible treatments that can help maintain its functioning.

Interstellar Travel

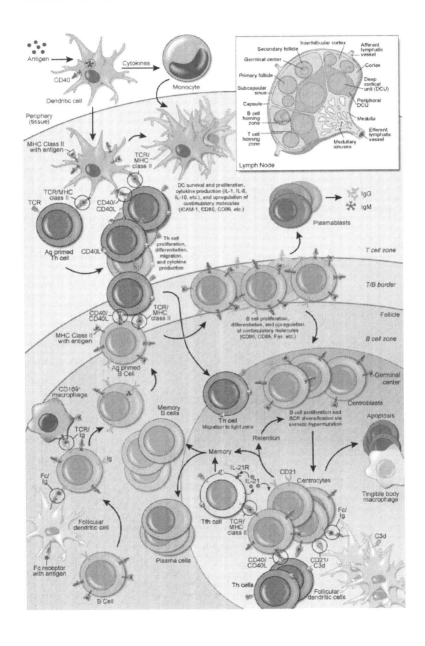

Physiological Effects

We have all heard about the discovery that astronauts exposed to weightlessness for prolonged periods develop a variety of physical problems. The human body is about 60% water, and lacking gravity, this water tends to concentrate in the chest and head. Legs and muscles atrophy, faces get puffy, and intracranial pressure increases. There is also the loss of bone density as gravity's stresses are removed and bones decalcify at about 1% per month. There are also problems with eating and sleeping. Astronauts get bored with menus and they often get only 5-6 hours of sleep each night instead of the recommended 8 hours.

There are also a new collection of problems that seem very perplexing and lack a known cause or treatment. In 2009 two medical astronauts, Dr. Michael Barratt and Dr. Robert Thirsk noticed that they were having problems seeing things close-up. They performed eye exams and confirmed this shift towards nearsightedness. They also saw evidence for swelling in their optic nerves and blemishes in their retinas. NASA sent up a high-resolution camera so that they could more accurately image their retinas, and they eventually found that the microgravity environment had physically altered the shapes of their eyes. The solution does not involve exercise, as was the treatment for very successful bone decalcification, but the actual reinstatement of near-Earth-normal gravity.

So there are many medical discoveries over the last 50 years that point to humans being more fragile in space than was previously thought. Apparently, it will take more that simple countermeasures to create an environment that is healthy for

Interstellar Travel

humans on long-term missions. We will need to supply them with almost constant artificial gravity to protect them against permanent eye damage.

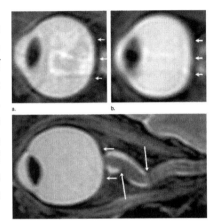

Top) Image of left eye before long-term exposure to microgravity. Note loss of convexity of the posterior scleral margin (arrows). Bottom) Image of right eye of different astronaut. Note two abruptly angulated foci (long arrows) in optic nerve sheath and posterior globe flattening (short arrows).

Artificial Gravity

Humans appear to be optimized for living under 1-G conditions. This can be relaxed a bit but not by more than 100% (zero to 2 Gs) before complications begin to arise. The most popular way to remove the physiological effects of space travel is to create artificial gravity, and the simplest way to do that is to rotate the crew quarters. In 1929 Herman Potočnik's *The Problem of Space Travel* was published, and he was the first to envision a "rotating wheel" space station to create artificial gravity. Since then, rotating toroidal stations have been the staple of science fiction. The reality, however, is that it is still technologically easier to build non-rotating stations like the ISS, and then to try to find medical interventions to solve the physiological problems like muscle atrophy and bone loss.

Physiological Effects

The physics of spinning stations is very simple, but there are some important issues that come from adding a human into the system. Humans have to move about, and depending on the size and rotation speed of the station, they will experience effects that lead to dizziness and nausea. These unpleasant effects can only be reduced by making the station larger and slowing it down, but this reduces its effectiveness in generating a useful gravity-like force. Here are the basic equations that let you design your own rotating space station with anti-gravity.

Imagine you are on the inside of a donut-shaped wheel. The geometry looks like the figure below. The distance from the axis to the midline of the torus is R and the radius of that inhabited station is r. The speed, S, of rotation depends on its 'RPM' or

Interstellar Travel

rotations per minute, and this formula is just $S = 2\pi R \times RPM$. If R is in meters, then the speed is in meters/minute. The amount of centrifugal acceleration you will feel is just $A = S^2/R$. Let's do an example.

Suppose the station (or the rotating part of the interstellar ship) has a radius of $R = 50$ meters, and it rotates at 5 RPM, then $S = 2 \times 3.14 \times 50 \times 5 = 1570$ meters/minute, or in our preferred units of meters/sec we just divide by 60 to get $S = 26$ meter/sec. The acceleration you will feel is just $A = (26)^2/50 = 13.5$ meters/sec². Since 1 Earth gravity is 9.8 meters/sec², this station is providing about $13.5/9.8 = 1.4$ Gs of artificial gravity.

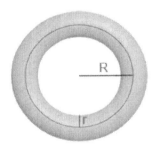

To get to a more comfortable 1 G, we can slow the rotation to about 4 RPM. We could also have chosen to decrease the radius R to 35 meters, or we could have adjusted both the RPM and the radius R to suit some engineering limitations.

Suppose the radius of the crew quarters, $r = 5$ meters. If we neglect the thickness of the walls, the volume of the torus is just $V = (2\pi R) \times (\pi r^2) = (2 \times 3.14 \times 50) \times (3.14 \times 5^2) = 25,000$ cubic meters. The habitable volume of the International Space Station is about 13,700 cubic meters, so this example is about twice as roomy as the ISS! Don't worry about all the other equipment. This is only the design for the part of the ship where the crew will eat, sleep and 'vacation' to maintain good health. Lots of other time will be spent in the 'weightless' portions of the ship.

Physiological Effects

This all sounds very nice, but here is a problem. You will only experience the floor as the outer half of the surface farthest from the axis. Your ceiling will be the surface of the torus closest to the axis. In this compact design, your feet would be a maximum of 50 meters from the central axis, but your head would be about 48 meters from the axis. While your feet are traveling with the 4 RPM rotation at a speed of 21 meters/second, your head will be moving around the axis at a speed appropriate to a radius R=48 meters and 4 RPM or 20 meters/sec. This also means that your head will feel an acceleration of 8.3 m/sec^2 or just 0.85 Gs. This doesn't happen on Earth, and the result is that your inner ear 'vestibular system' will not coordinate properly with its sense of where your feet are and what they are doing, and how you are visualizing the environment. Changes in any or all of these conditions cause disorientation, nausea and dizziness. Worse than this, as you move, you will feel the same kind of Coriolis force that hurricanes feel as they move due-north or due-south from the equator and are then deflected sideways. The only way to solve this problem is to make the radius of the torus, R, larger and/or reduce the rotation RPMs so that there is no longer a large acceleration difference between your head and feet. At some point in the design, the dizziness and vertigo problems will go away…hopefully!

NASA has undertaken many studies of rotating station designs over the years. Subjective Coriolis Illusion manifestations caused by disturbed space-station motion are considered the most detrimental to man's sense of well-being. The angular speed threshold is about 0.2 radians/sec. In our example of a toroid spinning at 4 RPM, we get 2π radians x 4RPM = 25 radians per

Interstellar Travel

minute or 0.4 radians/sec. This is above the threshold to trigger this nausea-and dizziness-producing effect. Notice that this doesn't rely on the actual radius or size of the torus, but only on its RPM!

So, to build an interstellar ship with artificial gravity, we have to keep its RPM below about 2 RPM. That means to get 1 G of artificial gravity, the torus must have a radius R no smaller than about 220 meters! This, by the way, is why massive space habitat designs like the ones by Gerard O'Neill are so huge.

Psychiatric Effects

Although physical and medical conditions can be treated presumably through chemical (drugs) or bio-engineering means, there is one last issue that remains a chief and well-understood concern for humans operating in space. Humans are subject to over 250 cataloged psychiatric conditions. Approximately one in four of us will suffer from one or more of these conditions in our lifetime. We commonly place a small number of humans in close-quarters, ask them to adhere to strenuous task schedules, and provide them a zero-gravity work/sleep environment, but almost predictably, this leads to a variety of psychiatric problems.

Between 2010 and 2011 the *Mars 500 Program* placed six men together in a simulated 520-day journey to Mars. The study found that when the crew was in charge of their own schedules that performed much better than when they were under constant contact with Ground Control and an external schedule over which they had little control. But there were changes in time perception, increases homogeneity in values, a reluctance to express negative interpersonal feelings, and increased feelings of loneliness.

In terms of selection, only a small number of candidates would be willing to be away from family and friends for the two- to three-year mission. This means that the pool of possible crew members will be restricted and possibly skewed psychologically in ways that cannot be foreseen. There will be a great deal of leisure time so boredom will be a challenge. Easy

Interstellar Travel

communication with family and friends back on Earth was a huge asset to maintaining proper psychological balance, but as the communication lags increase from seconds (ISS and Luna) to an hour or more, the sense of immediacy will be impaired and new adjustment problems will have to be developed. Even gazing at the huge disk of Earth from the ISS was a significant source of pleasure and connection to familiar experiences, but on journeys to Mars and beyond this will no longer be available. Will a telescopic view be enough?

Although NASA psychologists are confident that the ISS provides all of the necessary stimuli to keep the psychological health of astronauts in balance, they are not at all sure that long trips lasting hundreds of days or more to Mars and beyond will offer entirely new and unexpected challenges to overcome. Even the regular comforting sessions with psychiatrists back on Earth would therefore be impossible.

Psychiatric Effects

As astronauts spend longer times away from their friends and families back home, and the familiar conveniences of daily schedules and routines, feelings of isolation and aloneness will intensify. Disorders such as anxiety, post-traumatic stress, sleep loss/insomnia, adjustment, and depression can also develop unexpectedly in otherwise healthy individuals, even those who were previously not prone to these disorders.

Between March 1995 and June 1998, seven astronauts flew on the Russian space station *Mir* and during this time, psychiatric events were reported twice for an incidence rate for astronauts of 0.77 per person-year. Sensory perception can also be affected by changes in psychical state. In 1976 during the Soyuz- 21 mission to the Salyut-5 space station, the crew was brought home early after the cosmonauts complained of a pungent odor. Since the crew had not been getting along, the odor was considered to have been a hallucination.

Russian medical personnel view a condition called asthenia as one of the greatest problems affecting the emotional wellbeing of cosmonauts, defined as a nervous or mental weakness manifesting itself in tiredness and a quick loss of strength, low sensation threshold, extremely unstable moods, and sleep disturbance. This condition is particularly likely to occur when space flights last longer than four months.

Spaceflight also has positive effects on astronaut's mental health. For example, astronauts have reported having religious experiences, or feeling a sense of the unity of humankind while in space. The most frequently experience is the perception of the beauty and fragility of Earth. But during long-term missions,

Interstellar Travel

Earth may not even be visible, and this loss may have a negative impact on the well-being of astronauts.

Attempts at simulating long-duration missions with terrestrial analogs have provided mixed, but interesting, results. *Biosphere 2*, for instance, was an eight-member team isolated inside a 3-acre closed ecological system in Arizona for 2 years. Within 6 months, the team split into two factions where stolen food was horded, and daily tasks were reported as monotonous. By the end of the first year, some crew members became depressed. These effects were later credited to poor crew selection criteria, although similar behavioral problems have also been found in Antarctic 'winter-over' crews who remain isolated for up to two years under visually monotonous and confined conditions. Apparently, crew members who are rated low on the extrovert scale do much better than those who are 'normal' or high on this scale. In terms of career categories, polar crew members in the military rate higher than scientists on extroversion, and scientists rate lower than the support staff as well.

Another worry on long duration flights is the ever-present problem of boredom, which has been a major problem for polar explorers. Cultural and organizational affiliations can also contribute to the stress of space flight. One particular common instance of this is when ISS crews develop an "us vs. them" attitude between themselves and Ground Control. Psychologists call this "displacement" because the team is displacing the intra-group tension onto safer, more remote individuals.

NASA psychiatrists are pretty confident that with a small number of properly selected and trained crew members,

Psychiatric Effects

working in space for long periods of time can be such a pleasant experience that the crew may not even eager to come home! But this response, though common in the carefully-designed environment of the ISS, may not be possible under the far more extreme and remote conditions of deep space travel.

When men and women work together in space, as they have done in the ISS or in simulated environments on Earth, it is not uncommon to find instances of overt or implicit sexual stereotyping, which invariably leads to misunderstanding and tension. Conflicts among crew members based on differences in language competency and culturally-determined expectations, values, attitudes, and patterns of behavior have also been observed among long-duration Russian crews. However, congenial 'microcultures' do often form among scientists who share the same knowledge base, research experiences and outlooks on the tasks they are performing.

Companion Animals

One well-known approach to lessening some psychological problems is to provide companion animals. People residing in a long-term care facility such as a hospice or nursing home, experience health benefits from pets. Pets help them to cope with the emotional issues related to their illness. They also offer physical contact with another living creature, something that is often missing in an elder's life. A long-duration space voyage may offer similar restricted circumstances for which a pet might make a large, and perhaps crucial, psychological difference. They also can sense environmental changes of which humans may not be aware.

Interstellar Travel

Currently, only dead pets (one gram of their ashes) are allowed in space. A company called *Celestis Pets* lets pet owners can currently choose from four pet memorial spaceflight options. The cheapest is the Earth Rise service, which starts at $995. Pricier packages include the Earth Orbit ($4,995) and the Luna ($12,500). Also at the $12,500 price is the 'Voyager' package.

The big question is whether large animals like birds and other common animals would be able to adapt to microgravity. Birds would actually do very well in space because they can maneuver with their wings just as they do on the ground. However, free-flying birds do bring with them hygiene problems.

Twenty mice were brought to the ISS in September 2014 for a one month stay. On April 14, 2015, some 120 mice were brought to the station also to study microgravity effects on mammals. From previous flights, mice learn very quickly to grab on to the wire mesh in their habitats and crawl around on the walls and ceilings. Thus far, rodents are the largest animals that have been brought into space to temporarily live with humans. There are currently no plans for NASA missions to accommodate 'free-range' animals on future missions. But that doesn't stop science fiction writers for occasionally including pets in their stories.

Psychiatric Effects

One of the most famous pets in space is, of course, Jones from the 1979 movie *Alien* and the 1986 sequel *Aliens*!

Boredom

You are selected to be a crew member on one of the most exciting adventures ever conceived by humans: To travel to a distant star and survey or colonize a planet with a living biosphere.

Interstellar Travel

Your first day on the journey is filled with excitement about what you are doing and the fascinating ship environment. After the first week, you have gotten to know the layout of the ship very well, and you have mastered the intricate schedule of day-to-day activity and the tasks you have to perform. The view out the window is still intriguing, and with the ship's telescope you have dramatic views of Earth and the moon, plus the details of solar activity. Here we see an image taken by the NASA Cassini spacecraft showing Earth and moon from a distance of 1 billion kilometers. The round-trip radio delay is now just under 2 hours.

Months later, there is not a single nook and cranny in the cavernous volume of the ship that you haven't visited numerous times. The Earth and sun are now just unresolved dots of light even through the telescope. Communication delays with Earth have now lengthened to a dozen hours and you have had to completely adjust to loosing synch with your loved ones. Your daily tasks are endlessly repetitive and it is impossible to feel much joy in them now that the newness of your adventure has

Psychiatric Effects

finally worn off. You have already been outside the ship several times for routine inspections and even that activity is starting to get annoying. Other than the articulated surface of the ship, the view of the sky is monotonous now that the sun and solar system have vanished into the sameness of the stellar universe. All of humanity has collapsed into a single bright star in the constellation of Gemini, and the journey still has a mind-numbing dozens of years to go.

At some point you and the other members of the crew will confront the sameness of your environment and schedule and fall victim to the psychological state known as boredom. This is not a state to trifle with because it is the way that our brain reacts to insufficient stimulus.

"Boredom, it turns out, is a form of stress. Psychologically, it's the mirror image of having too much work to do, says Jason Kring, president of the Society of Human Performance in Extreme Environments, an organization that studies how people live and work in space, underwater, on mountaintops and other high-risk places. *If your brain does not receive sufficient stimulus, it might find something else to do—it daydreams, it wanders, it thinks about itself. If this goes on too long, it can affect your mind's normal functioning. Chronic boredom correlates with depression and attention deficits."*

This typical scenery on Mars will be fascinating for about the first few days. What will you think of it in a week? A month? A year? Try not to 'romanticize' how you think you will feel as an adventurer. Unless you are a geologist or someone else who can receive daily stimulation about it, everyone else will see it for the

Interstellar Travel

monotonous, mono-color scenery that a dispassionate eye instantly appreciates.

You have experienced many of defenses against boredom yourself, and if this condition persists it starts to seriously affect the way your brain and you work, leading to depression and attention deficits. What if a chronically distracted crew member fails to make a critical measurement accurately, or gets confused about a routine safety procedure? Boredom also correlates with increased risk-taking as you try to change a boring routine into something more stimulating, and potentially more hazardous just for the momentary trill of it. Throwing rocks on Mars sounds fun, unless your colleague happens to be on the receiving end! You might consider taking spacewalks more frequently, which uses up resources, or perform the spacewalks without a tether.

Psychiatric Effects

Here we see astronaut Bruce McCandless drifting above the changing face of Earth, but on our deep-space journey, only the spacecraft will be our familiar local scenery. All else will be pinpoints of light in the sky.

Perhaps the best-studied people subject to boredom are residents of Antarctic research stations, and astronauts on board the International Space Station. In fact, ISS astronauts rarely complain of boredom because their daily schedules are jam-packed with work. Although the interior of the station is a near-hazardous jumble of cables and crowded wall space, the view of Earth outside the many windows is constantly changing from minute to minute along the orbit.

Contrast this with McMurdo Station, where the view outside is usually featureless white, and for month after month your cramped habitat rarely changes, nor your schedule. To counteract boredom, Antarctic crews perform skits and plays, or celebrate many holidays both real and invented. Alcohol consumption goes up dramatically. NASA is very concerned about boredom on even comparatively short 8-month flights to Mars. Next to the hazard of radiation exposure, boredom is considered the next major factor that could pose a severe risk to the crew and the completion of the mission.

Astronaut Suni Williams spent six months on the ISS. She made the most of her time by running the 26-mile Boston Marathon on a treadmill, and followed her beloved Red Sox closely. She told *ABC News* that she just pretended she was on a long camping trip. Williams also played several games with Mission Control. She would toss out song lyrics and make flight

Interstellar Travel

controllers guess the song. She also kept a log of the places over which the shuttle flew.

But for the journey to Mars, round-trip time delays of up to 25 minutes will make direct communication with family and friends impossible in the normal conversational sense. There will be little to see outside the window except empty, star-filled space. And a half-dozen crew members jammed into a volume about the size of a small summer cottage will bring out many psychological states over time. From studies of Mars simulation experiments *Mars500* and *HI-SEAS* (Hawai'i Space Exploration Analog and Simulation), the effects of situational boredom and food monotony were studied in detail. Crews kept diaries and were examined by psychologists to see how behaviors evolved over time. Some crew members were able to adapt to the monotony by using extensive music collections, reading, playing complex games, and finding ways to vary their daily schedule, while other crew members seemed less adept at coping and became depressed and withdrawn. The bottom line was that environment mattered, as did communication with family and friends, but for a Mars mission the experience will be vastly different than experienced for a similar time period on the ISS or even McMurdo Station. Psychologists are prepared to learn about new kinds of boredom states requiring even new coping strategies. For interstellar journeys, or even multi-year trips to distant solar system outposts, it is fully expected that vastly new coping strategies will have to come into play to insure the psychological health of the crew. What those strategies will be, is a complete mystery today.

Psychiatric Effects

The famous movie sequence and music from *2001: A Space Odyssey* of astronaut Bowman exercising within an monotonous white environment on a many-month flight to Jupiter, might hold some clue. HAL did all the heavy-lifting of keeping the ship operational. This suggests what long-duration flight with an

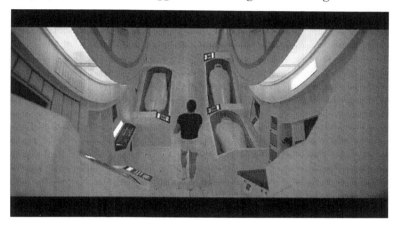

artificial intelligence system might be like. Humans will have to take a far more active role in maintaining the ship and not let automatic systems have all the fun. AI systems may only be needed to 'watch the astronaut's back' in case they miss something vital going on.

Will crew members turn to narcotics secretly created from ship's medical pharmacy? Will graffiti start showing up on otherwise featureless wall space? Will these be seen as crimes of vandalism, or merely normal crew behavior not to be confronted until it becomes a hazard?

It is hard to imagine anything more beautiful or stimulating than artwork replacing the mind-numbing, white or battleship-gray

Interstellar Travel

wall space. However, if the ISS is any example, there may not be any free wall space to decorate that is not already taken up by vital equipment, cabling, hoses and other technological necessities.

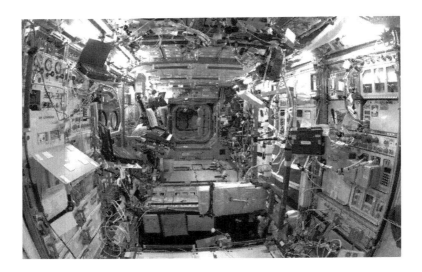

Us versus Them

The most damaging behavior for group behavior is the Us versus Them idea. This produces a variety of social and psychiatric stresses as groups of people sort themselves by physical attributes, philosophical or religious ideas, and even the valuation of secrecy where one faction knows something that the other does not. Unfortunately, it seems to be hard-wired into our primate brains. It's a gift from the time when millions of years ago we lived on the Savannah. According to research in behavioral neurology and cognitive brain science, we needed to develop a keen sense of our immediate group 'us' and everyone else 'them'. We also needed a Leader to make snap decisions based on immediate impressions and not on long deliberative discussions. Those who did this best survived another generation against predators. This also impacted social cohesion by insuring that the 'in group' developed a set of core ideas and survival behaviors that tended not to be questioned. This in turn led to the creation of laws that promoted certain behaviors and discouraged others. In fact, the collection of laws and beliefs became over time the principle way in which one group distinguished itself from another, other than by obvious physical characteristics.

It is an interesting question whether there exist collective social behaviors that could be a liability for a group attempting to survive cut off from the rest of humanity for an extended period of time. For short 'interplanetary' trips, there is always the assurance that you will rejoin humanity at the end of the journey, but for interstellar travels, there is no such guarantee.

Interstellar Travel

The simplistic solution is to make sure that the crew is as homogenous as possible in terms of culture, ethnicity, race, religions belief and so on. Of course, one of the great strengths of humanity is the way it has used diverse attributes and opinions to advance, solve problems, learn adaptation, and the value of negotiation and bartering. Nevertheless, human history does not show that we have a good track record for tolerance, especially when it involves religious beliefs, or especially resolving the tensions between 'Us' versus 'Them'.

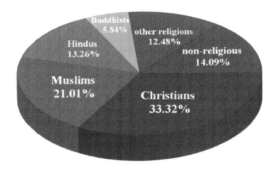
World Religions by percentage

All religions have their examples of innocents slaughtered over ideology, and it is a valid concern whether the crew will be able to practice a diverse set of religious beliefs, or whether humanity would even allow crew selection based either on non-theistic beliefs, or the crew representing the diverse views of humanity, of which 95% are theists.

In classic Us vs Them confrontations, religions have a limited history of peaceably coexisting with other belief systems. Christians and Muslims have historically demonstrated the ability to create from within their own theological fabric,

Us versus Them

differences of opinion that evolve into separate groups be they Protestant vs Catholic, or Sunni vs Shia. It is also true that even the most cooperative religions in the short term can evolve over time ultra-orthodox factions that would have a devastating effect on the free exchange of information and opinions critical to the success interstellar trip lasting decades or even centuries. Even the very concept of such a voyage may find itself at odds with a particular theological doctrine. One thing that will work against this, however, is the certain knowledge that their collective survival depends on openness, communication and cooperation.

In 2014, University of Dayton political science assistant professor, Joshua Ambrosius, used data from the General Social Survey and three Pew surveys to compare knowledge, interest and support for space exploration among Catholics, Evangelicals, Mainline Protestants, Jews, Eastern religions and those with no religion. What he found was that Evangelicals, who account for 25% of the US Population, are the least knowledgeable, interested and supportive of space exploration, while Jews and members of Eastern traditions were most attentive and supportive. Regular church attendance, a literal view of the authority of the Bible, and belief in creationism, appear to exert a negative effect on support for space exploration. Would such an anti-science viewpoint be consistent with being a positive member of a crew? Your religious beliefs definitely shape how you view science and space exploration. So the logical question would be whether Evangelicals would be suitable crew members? How would the human population that will fund this massive effort regard some beliefs being identified as prohibited on such a flight?

Interstellar Travel

In February 2014 the General Authority of Islamic Affairs and Endowment (GAIAE) issued a fatwa forbidding devout Muslims from participating as crew members in Mars Ones proposed one way mission to Mars. Speaking for the clerical group, Dr Farooq Hamada explained that, "*Protecting life against all possible dangers and keeping it safe is an issue agreed upon by all religions and is clearly stipulated in verse 4/29 of the Holy Quran: Do not kill yourselves or one another. Indeed, Allah is to you ever Merciful.*"

The Pew Research Center reported in 2009 that 61 percent of the American public said science doesn't conflict with their religious beliefs. And 52 percent of people who attend church weekly also don't believe new scientific knowledge conflict with their beliefs, according to the study. Nevertheless, the risk of accidentally inviting just one clandestine fundamentalist into the crew selection could have enormous negative consequences.

Leading Christian creationist Ken Ham says space exploration is driven by man's rebellion against God, and that if aliens do exist, they are going to hell. Ham, writing a for his *Answers in Genesis* website, claims that the U.S. space program is a waste of money because any alien life that scientists might find would face eternal damnation. Ham's post is a rant against "secularists" and their need to "prove evolution." Ham explains that space exploration is "rebellion" against his God. Even if extraterrestrial life forms do exist, they would be stained by original sin as committed by Adam. However, because the aliens would not be descended from Adam, they would not be entitled to the imagined "sweet salvation" waiting for the Christian.

Us versus Them

Religious leaders attending the *100 Year Starship Symposium* - a meeting to discuss the prospect of sending a space mission to another star within 100 years, tended to agree with the comment by Baptist Rev. Alvin Carpenter: *"The only way humanity can survive is if they leave behind the Earth-based religions...If there's any way to make this fail, bring Earth-bound religions."* Religions, he argued, breed aggression and conflict, citing the violent history of his own faith, Christianity, in episodes such as the Inquisition and the Crusades. Many religions' negative stance on homosexuality has driven young gay people to commit suicide, he said. *"Where humans go, they take religion with them,"* he said. *"Even if we screen people and we say no religious principles, 100 years from now people are still going to return back to those things."*

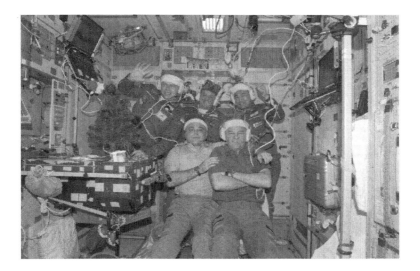

Astronauts celebrated Christmas on the ISS in December, 2009. But how would Hindis, Muslims and Jews feel about this seemingly harmless celebration of an important holiday not of their faith and tradition? How will Muslims pray five times a day

Interstellar Travel

facing Mecca...in outer space? Will facing Earth be enough, or will it be sacrilegious because the direction also includes other geographic localities as well?

Is it reasonable to expect groups that have had thousands of years of bitter history to step beyond these prejudices and hurtful, historical baggage, to suddenly become the ideal, cooperative astronaut? Although no such issues have emerged from astronauts onboard the ISS, expeditions to Mars will be the first long-duration test of cooperation and necessity triumphing over latent prejudices and misunderstanding.

Because it's the Law!

What laws will we bring with us and how will we inforce them? Will we have money?

Depending on the size of the crew, and especially the size of the conscious crew in the case of large-scale hibernation spacecraft, the concept of a law will likely change. Because the environment will be unlike what was left behind on Earth, many if not most laws we might bring with us from Earth will be meaningless. There will also be no meaning to laws that place monetary value on objects, if the crew and colonists live in an environment not based upon assigning monetary value to work, productivity or goods.

Once you have laws, you will need a system of enforcement, penalties, police, incarceration space, a penal/judicial system, as well as advocates for the accused. Beyond this, you will need some form of governance where votes are tallied and decisions made on the basis of some system that insures the will of the majority is carried out without delay, but where the desires of the minority are also protected and respected. Unfortunately, once a vote is taken, you immediately have an Us versus Them situation, and all of the long-term issues that that can bring out. When individuals are repeatedly among the Others because their wishes consistently get outvoted, alienation can occur and that could lead to hostility and even law-breaking.

For a closed environment like an interstellar spacecraft, all actions must be evaluated on the basis of their ability to do harm to the entire group and mission. Behavior that leads to the

Interstellar Travel

development of hazards, both to the flight hardware and to crew members compromising either the hardware or human health, will have to be scrupulously restricted. The question of whether private guns or other projectile devices will even be allowed on the ship at its launch is a significant issue.

As for the tendency to be violent or to engage in law-breaking, world statics seem to show that some countries have decidedly lower per-capita crime rates than others. We may well decide to select crews partly on the basis of the kinds of cultures they have been exposed to, rather than open the selection up to all nations on an equal basis. Consider how different countries rank in crime statistics shown, for example, in this diagram.

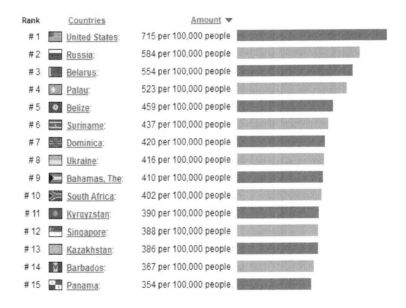

Rank	Countries	Amount
# 1	United States:	715 per 100,000 people
# 2	Russia:	584 per 100,000 people
# 3	Belarus:	554 per 100,000 people
# 4	Palau:	523 per 100,000 people
# 5	Belize:	459 per 100,000 people
# 6	Suriname:	437 per 100,000 people
# 7	Dominica:	420 per 100,000 people
# 8	Ukraine:	416 per 100,000 people
# 9	Bahamas, The:	410 per 100,000 people
# 10	South Africa:	402 per 100,000 people
# 11	Kyrgyzstan:	390 per 100,000 people
# 12	Singapore:	388 per 100,000 people
# 13	Kazakhstan:	386 per 100,000 people
# 14	Barbados:	367 per 100,000 people
# 15	Panama:	354 per 100,000 people

Because it's the Law!

Another issue is, what form of authority structure would work best in interstellar voyages? For decades, we have seen science fiction writers offer only one model based on a ship's Captain and the usual crew structures, with everyone else considered a passenger. This is also the foundational structure of the original *Star Trek* series and the newer *Star Trek:The Next Generation*. A military command structure was also implied in the old *Outer Limits* episodes of the early-1960s.

Today, the International Space Station is also based on a strict adherence to a naval command structure, supported by a ground-based Mission Control system. But a reasonable question to ask is, what is the evidence that this kind of top-down structure applies to long-duration missions. Is there something in human behavior that is, say, genetically pre-programmed that makes these kinds of systems a good idea regardless of circumstances?

In isolated and confined environments, two important leadership roles have been identified: the task/instrumental role (which focuses on work goals and operational needs), and the supportive/expressive role (which focuses on morale goals and emotional needs). On the International Space Station, the mission commander should be familiar with both of these aspects of leadership. For crewmembers, the supportive role of the commander (but not the task role) related positively with crew cohesion. For mission control personnel on the ground, both the task and supportive roles of their leader were related positively to mission control cohesion.

Interstellar Travel

The Code of Conduct for International Space Station crews establishes a clear chain of command and relationship between ground and on-orbit management, standards for work, responsibilities with respect to elements and equipment, disciplinary regulations, together with physical and information security guidelines. It also defines the ISS Commander's authority and responsibility to enforce safety procedures, physical and information security procedures and crew rescue procedures. In cases where necessary to ensure the immediate safety of the Crew Members of the ISS, reasonable and necessary means may include the use by the ISS Commander of proportional physical force or restraint'. Cooperating Agencies understood upon signing on to ISS participation that force would be used only when immediate safety was jeopardized, and after exhaustion of other possibilities.

This kind of system with one Leader and others working together on their separate tasks guided only by an agreed-upon code of conduct, works well when you have a small number of crew members up to the size of a football team, with the quarterback calling the plays, and your various team member contracts stipulating and requiring agreement to this command structure with financial penalties levied for infractions, but what happens if you have larger crews? At what point does a sole Captain-led model with everyone else a roughly equal subordinate begin to break down? Also, how are penalties and the authority of Earth weighed against the requirement of a near-term punishment for infractions to correct the immediate behavior?

Because it's the Law!

The US Navy has always provided a model for how to manage crews of up to a few thousand members. It is very authoritarian, provides local policing and punishment (the infamous Brigg), and begins with the civilian President of the United States as the Commander-in-Chief. On interstellar voyages, there would be no return to harbor, and the ship's leader or Captain would be the ultimate law-giving authority.

The naval chain of command looks like: Captain – Commander – Lieutenant Commander – Lieutenant and finally the Ensign and Midshipman. The level and complexity of leadership training increases steadily from Midshipman to Captain rank. These are called the Commissioned Officers. Under Ensigns you have the non-commissioned crew consisting of Chief Petti Officers (Sergeant Major), Petty Officer (Sergeant), Leading Seaman (Corporal) and Seaman (Private). The role of the non-commissioned crew is still bound by the military command structure and the Uniform Code of Military Justice.

The research ships at NOAA typically have crews of 20-30 people, and are run by a combination of NOAA Commissioned Officers and Wage Marine civilians. The Naval Officers follow a typical naval chain of command, though will fewer positions. The ship's commissioned officers and crew provide mission support and assistance to scientists from various NOAA laboratories as well as the academic community.

The Wage Marine personnel include licensed masters, mates, engineers, and unlicensed members of the engine, stewards, and deck departments. In addition, survey and electronic technicians operate and maintain the ship's mission, communication and

Interstellar Travel

navigation equipment. They are grouped to form the various ships departments such as Engineering, Deck, Fisherman, Survey and Steward (cooks); each have their own internal ranks such as Chief Engineer, Assistant Engineer, Junior Engineer, Engine Utilityman. These ranks are generally decided based on technical competence and seniority at sea. The leadership is less formal and more akin to academic systems based on years of experiences with Principle Investigators at the top, followed by co-Investigators, post-doctoral researchers, graduate students and undergraduate students.

Through a process of social evolution over millennia, humans have developed leadership chains as an effective and efficient means of making decisions under times of survival, where time is of the essence. In these critical situations 'too many cooks in the kitchen' can prevent critical decisions to be made quickly. Leadership structures are found in as few as groups of two or three people! There is definitely a genetic basis for creating a good leader. About 48-59% of leadership qualities seem to have a hereditary basis according to a statistical study performed by researchers at the National Institutes of Health in 1998.

The very make up of our brain is also predisposed to people having a variety of leadership skills that seem to be ultimately based on the so-called 'fight or flight response'. This behavior suited us very well when we were confronted by a saber-toothed tiger, but was also co-opted to serve other needs found in social groups. When you have a set of needs, do you challenge someone, negotiate, or avoid confrontation? A brain region known as the amygdala controls this response and our higher-

Because it's the Law!

level brain functions have to figure out what to do with the resulting behavior in a socially-appropriate way.

So the need for leadership structures is not only something hard-wired into our brains, but is genetically transmitted, and so for long-term spaceflight we have to recognize that some kind of leadership hierarchy will have to be set up. It is an intimate part of who we are as humans.

Someone always wants to be The Leader.

As crews get larger, duties become more compartmentalized, leading to a separation of groups who make decisions for the safety and functioning of the ship, from those members responsible for the scientific research and instrument maintenance. The model we select for an interstellar voyage will depend on the size of the crew we send, and the minimum number of people needed to insure the success of the mission.

Interstellar Travel

Here is a typical crew of the ISS. A similar-sized crew is contemplated for our first journey to Mars in the 2030s. In this system, there is only a slight difference between the tasks assigned to the Commander and that of the scientists and engineers in the Crew.

Here is the crew of the NOAA vessel Pisces. Can you identify the commissioned officers? The crew? The lone scientist?

What's for dinner?

Around the world, only 5% of the population claims to be vegetarian. The rest enjoy a piece of meat, poultry or fish on a regular if not daily basis. Meat, as you may know, is very expensive to produce in the traditional way.

Animal protein production requires more than eight times as much fossil-fuel energy than production of plant protein, while yielding animal protein that is only 40% more nutritious for humans than the comparable amount of plant protein. Chicken meat production consumes energy in a 4:1 ratio to protein output; beef cattle production requires an energy input to protein output ratio of 54:1.

Grain-fed beef production in the United States takes 100,000 liters of water for every kilogram of food. Raising broiler chickens takes 3,500 liters of water to make a kilogram of meat. In comparison, soybean production uses 2,000 liters for kilogram of food produced; rice, 1,912; wheat, 900; and potatoes, 500 liters.

Americans eat about the same amount of meat as we have for some time, about eight ounces a day, roughly twice the global average. Americans are downing close to 200 pounds of meat, poultry and fish per capita per year. It's likely that most of us would do just fine on around 30 grams of protein a day, virtually all of it from plant sources.

Interstellar Travel

Sugar - In 1822 the average American consumed about 45g of sugar every five days (the amount in a 12oz can of soda). Today, that number is 756g!

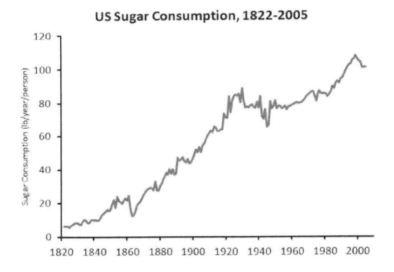

As the chart shows, we're now eating about 100 pounds a year apiece. Sugar has been linked to obesity, and suspected of, or fully implicated as a cause in the occurrence of diabetes, cardiovascular disease, dementia, macular degeneration, and tooth decay. Numerous studies have been undertaken to try to clarify the position, but with varying results, mainly because of the difficulty of finding populations for use as controls that do not consume or are largely free of any sugar consumption.

Chocolate - Regular consumption of chocolate could weaken bone density and strength, which could in turn increase the risk

What's for dinner?

of osteoporosis. According to a new study, women who eat chocolate daily have an overall bone density 3.1 percent lower than those who consume it less than once a week. More than 1,000 women aged 70-85 were asked to keep a diary of how often they consumed chocolate or cocoa-based drinks.

Plants - Research on the International Space Station suggests plants can grow in microgravity, but require light levels far higher than typical human indoor levels. The good news is that plants grow just find at 1/10 the normal Earth atmospheric pressure, which greatly relieves the technology of a fully-pressurized greenhouse. New studies of greenhouses connected to human quarters suggests that enough plant oxygen would be generated in a few weeks to become a fire hazard.

A typically overlooked requirement in most space trips is the medium in which plants will be grown. On Earth this is commonly soil, but hydroponic systems are also used. The problem is that soil is a whole ecosystem, containing bacteria, protozoa, nematodes, insects, and much more—and it supports us in many ways, in particular, it is important for maintaining the proper diversity and balance of microbiota (i.e., bacteria) in the human gut. Martian colonists could probably live for years on food grown without soil. The question is, could they live on it for decades? Could their children grow up on it?

Microbiologists are linking decreases in gut biodiversity to Western diseases such as allergies, asthma, irritable bowel syndrome, and colon cancer. Permanent residents of Mars, and their Mars-born children, could be even more afflicted with

Interstellar Travel

these problems than people on Earth. Mars may be both too dirty—in the sense of having toxic dust—and too clean.

Simply put, humans need good dirt. The human body harbors 100 trillion bacteria inside and out, and their proper balance is increasingly regarded as vital to human health. Just getting to Mars could throw the bacterial balance of arriving colonists out of whack. Space does odd things to bacteria. According to Hernan Lorenzi, a biologist at the J. Craig Venter Institute, nasty bacteria like salmonella and pseudomonas become even more virulent when grown in Petri dishes in weightlessness.

Forty years of research has shown that astronauts' gut bacteria change during long missions. Until those problems are solved, the best solution may be to send food from Earth to supplement what the colonists manage to grow on Mars. The idea is to get the colonists eating a non-Western diet, with lots of vegetables and fermented foods like yogurt, sauerkraut, kefir, and kimchi. Such foods are laden with healthy bacteria. Customized probiotic pills might also help, if they can be shown to work.

MarsOne has stated that they will be sending crews of four to Mars to create a permanent settlement on Mars each year, beginning in 2024. The founder of *Space X*, Elon Musk, speculates that the colony will start out vegan due to the space and energy it takes to raise animals for food. It turns out there are some important reasons why it should stay that way.

By making the Mars colony a permanently vegan, plant-based society, we will be cutting out the need for beef and poultry

What's for dinner?

factory farming. This would reduce energy costs, conserve space, reduce toxic waste, avoid carbon dioxide, methane, and nitrous oxide air pollution, reduce the speed at which resistant bacteria mutate, reduce long-term medical costs, and preserve water and food for human consumption. It would also significantly reduce their risk for many diseases and conditions, improve their overall heath, bolster their immune systems, accelerate healing should any accidents occur, and keep their bodies and minds in optimal condition for the task at hand. For all these reasons, veganism is a win-win for both the humans and their environment in this new colony.

Growing plants requires a lot less resources than growing meat needs. Plants also emit life-giving oxygen rather than especially toxic feces. In fact, we now have the technology to produce huge amounts of plant food, like lettuce, in a very efficient way. According to research by the American Dietetic Association, humans actually have no nutritional or biological need for animal products. In fact, a vegan diet is reported to be much

Interstellar Travel

healthier than a non-vegan one. The human digestive tract actually evolved to digest plant food better than anything else. Carnivores have very short gastrointestinal tracts in order to expedite the removal of decaying meat from the body, whereas the human digestive system is long enough to optimally digest fibrous vegetation. Carnivores also have special features designed for digesting meat, such as highly acidic stomach acid. On the other hand, humans have special features for digesting different kinds of plant matter, such as the enzymes in the saliva that begin the digestion of starches. Because our bodies evolved to handle plants best, we are well nourished when they are the basis of our diet and have adverse health effects when they are not.

Humans can receive all eight essential amino acids in their diet as long as they eat unprocessed starches like beans, potatoes, corn, or rice, as well as fruits and vegetables. Other sources of plant protein include whole grains, nuts, and seeds. The only essential vitamin that plants lack is vitamin B12. B12 is made by the fermentation of certain bacteria. Cows, for example, get their B12 from the bacteria that ferment during the first stage of digestion. Scientists are able to produce B12 a similar way. This means that humans can receive their full B12 requirements with a simple supplement.

Printable Food - 3-D printing certainly isn't new to the space technology arena. It has been pegged as a potential answer to building habitats on the moon and Mars, for example. A 3-D printer could be landed onto a planetary body ahead of a human mission and it could automatically construct habitable structures out of locally-mined materials.

What's for dinner?

This image shows a printer delivering chocolate to a cracker. At least it's a start. To Anjan Contractor, of Systems and Materials Research Corporation (SMRC) in Austin, Texas, food is just a collection of compounds that can be isolated in the form of a powder. These different varieties of powder can then be mixed with oils and water, combined by his 3-D printer. Governed not by a paperback recipe book, the printer would be controlled by software recipes. Food can then be printed to order.

Although the applications in space are obvious — and even Jean Luc Picard would be impressed by this modest attempt at a Star Trek replicator — Contractor is acutely aware that this idea may, one day, transform the way we conceive food here on Earth. This in-turn may help feed the planet. *"I think, and many economists think, that current food systems can't supply 12 billion people sufficiently,"* he said. *"So we eventually have to change our perception of what we see as food".*

As the picture shows, 3-D printing is now a rapidly growing reality in the candy business! It's all a matter of what you extrude, and the raw ingredients are basic starches, fats and carbohydrates with interesting flavoring mixed-in to confound and stimulate our sense of taste and texture.

Interstellar Travel

Funny Odors

Imagine being cooped up in a closed environment with a few or a dozen other humans going about their daily routines. Invariably there will be odors that are generated by outgassing from electronic devices, food odors, odors from normal flatulence, and from toilet activity. The standard approach is to use filtration to reduce or remove these odors, and this seems like a workable response. But it's not so simple.

On the ISS, over two hundred aromatic contaminants have been identified and designed for. The extensive air purification system has been deemed very effective by astronauts, even as it deals with sweaty work-out garments for the mandatory 2-hour sessions every day. Bathroom odors are seemingly never a problem, however some potent foods like American seafood gumbo are all but forbidden on the ISS due to their almost universally disliked aroma. The Trace Contaminant Control Facility takes in air from the ISS, heats it to 400 C to reduce all organic components to carbon dioxide and water using a catalyst. The non-organic materials are converted into acidic gases and then removed by charcoal filtering. The gas is then shilled to room temperature and returned to the ISS cabin.

But although the end result is considered odorless by astronauts, there is another odor that cannot be identified other than to describe it as the "smell of space." According to Astronaut Clayton Anderson, *"Oft mentioned by astronauts, the smell of space is somewhat hard to describe. Ever distinct—I would know it instantly if I smelled it—it has been likened to smells associated with welding or burning*

Interstellar Travel

of ozone (now who the heck really knows what that smells like?!). Most noticeable following a spacewalk, when crews and their equipment returned to the inside of the ISS, I remember being able to smell traces of this unique scent for several days following an excursion into the unforgiving vacuum of space."

According to Astronaut Don Pettit: *I had the pleasure of operating the airlock for two of my crewmates while they went on several space walks. Each time, when I repressed the airlock, opened the hatch and welcomed two tired workers inside, a peculiar odor tickled my olfactory senses. At first I couldn't quite place it. It must have come from the air ducts that re-pressed the compartment. Then I noticed that this smell was on their suit, helmet, gloves, and tools. It was more pronounced on fabrics than on metal or plastic surfaces. It is hard to describe this smell; it is definitely not the olfactory equivalent to describing the palette sensations of some new food as "tastes like chicken." The best description I can come up with is metallic; a rather pleasant sweet metallic sensation. It reminded me of my college summers where I labored for many hours with an arc welding torch repairing heavy equipment for a small logging outfit. It reminded me of pleasant sweet smelling welding fumes. That is the smell of space.*

When an astronaut spends enough time in a zero-gravity environment, they get what NASA calls the 'Charlie Brown' phase. Their head begins to swell, because liquids that are usually dragged down by gravity stay in their head. As a result of this, astronauts begin to lose the sense of smell. They feel like they have a cold, stuffy nose, etc. And they can't really taste things very well. The head of NASA's food science program thinks that losing the ability to taste leads to astronauts craving hot things like Tabasco because that way they can at least feel something different in their taste buds.

Funny Odors

Not much is known about the long-term effects of prolonged stays in space. Studies on the matter are coming out more and more. However, previous research has shown that reduction of the sense of smell leads to greater risks of depression; an added problem for long-duration manned space missions.

By far, for ISS astronauts, the most popular foods are shrimp cocktail and Tabasco sauce. Perhaps the most vocal fan of space shrimp was astronaut Story Musgrave. The six-time shuttle flier was known to eat the stuff for every meal, even for breakfast. Under the plans for a Mars mission, astronauts would eat freeze-dried food on the 300-day journey, but once there they could attempt to grow their own fresh food. Interstellar travelers may demand more than this level of culinary and sensory stimulation!

In the 1990s, NASA scientists were thinking of what astronauts would need to survive long-term missions. One important requirement was a dependable source of food. But cultivating crops in a sealed-off environment results in the buildup of an undesirable gas called ethylene. Plants release the odorless, colorless fume into the air, which has the unfortunate effect of accelerating decay, hastening the wilting of flowers and the ripening of fruits and vegetables. To address the problem, NASA developed plant growth chambers that included an ethylene "scrubber," where air is drawn into tubes coated with thin layers of titanium dioxide. When an ultraviolet light source located in the tubes strikes the coatings, the ethylene gets converted to water and carbon dioxide, both of which are beneficial for plants.

Interstellar Travel

Diurnal cycles

Also called the circadian rhythm, it is an internal clock that regulates key metabolic processes as well as mental alertness and performance. For most complex biological systems on Earth it is 24-hours long and is regulated by the light-dark cycle throughout the Earth day. Even animals that live in caves have primitive photoreceptors that sense daylight and help reset their diurnal clocks.

Humans get their reset key from light falling on the retina, which is passed on to the pineal gland, which then secretes melatonin. Secretion of melatonin peaks at night and ebbs during the day and its presence provides information about

Interstellar Travel

night-length. This 24-hour beat is not immutable, and we can actually do just fine with periods from 23.5 to 24.7 hours. Disruption to rhythms usually has a negative effect, which can include fatigue, insomnia, disorientation as well as bipolar disorder. Long-term disruptions are especially adverse and can cause cardiovascular disease and renal failure.

NASA astronaut studies show that the circadian rhythm of astronauts in the ISS are similar to what they have on the ground. However, in-flight sleep showed a decreased amount of sleep obtained to about 6 hours, and astronauts showed a decrease in delta sleep. To compensate, in some cases nearly half of all astronauts take sleeping pills.

After a few weeks of this reduced sleep, astronauts perform at the level of someone who has been awake for 24 hours straight. The resulting fatigue is a serious problem for astronauts who have very busy schedules of activity requiring high levels of attention and higher-level thinking and problem solving. It also affects performance, increasing irritability, diminishing concentration, and decreasing reaction time as well as increasing the risk of accidents.

Beyond performance, disruptions to the circadian rhythm may affect physical and mental health. Sleep disorders, cardiovascular disease, diabetes, obesity, cancers, inflammatory disorders, and mood disorders (depression, schizophrenia, and attention deficit) may result. One effective cure that seems to reset the circadian rhythm is exposure to very bright light for 45 minutes every few days. In the ISS the lighting is generally 100 to 500 lux and not the 2500 lux needed to perform this function.

Diurnal Cycles

Without the periodic circadian rhythm reset, severe difficulties usually occur. The classic example is what happened to Earth-based operators of the Mars Pathfinder rover. The rover was solar powered, operated on Mars Time, which is a day about 40 minutes longer than on Earth. Ground controllers on Earth were forced to live according to a Martian day/night cycle. The team's work schedule floated through two time zones every three days. The team creeps from day shifts to night shifts and back. After a few months of this, ground controllers became uncooperative and even by some accounts mutinous.

Interstellar Travel

Communication

The math is pretty simple, but the consequences are enormous.

At the speed of light, communication delays are given by distances in light years. A star 25 light years away will take 25 years to receive a signal from Earth, and 25 years to get a reply. Back on Earth, 50 years will elapse between each contact, and you need to ask whether a future Earth civilization can carry on this kind of delayed discussion in a meaningful way. If our first target is an Earth-sized planet with a biosphere 100 light years away as the statistics might suggest, the delays become centuries. What would be the nature of these transmissions?

If the Earth signal were transmitted in 1800 AD during the industrial Steam Age and before the Age of Electricity, it would reach the travelers in 1900 AD when travelers would hear about the pre-electricity world of humankind. They would send a reply back to Earth, which would arrive in 2000 AD, and the travelers would get Earth's response in 2100 AD. The traveler's second contact with Earth will now be discussing interplanetary travel and intelligent computers, not steam-powered locomotives! How could you and Earth possibly keep up with the changes in context, and the evolving needs of this communication channel? In 200 years, none of the traveler's families will have survived a 10-generation dislocation. You would share genes with these people, but all sense of common familial or even social history will have vanished.

Interstellar Travel

Clearly, the strategy would be to continuously broadcast all manner of information, not just 'Hi how are you' and 'Doing fine!' The information arriving at its destination would flow continuously over the decades or centuries giving a continuous record of what's going on Back Home, but there would be few if any personal exchanges to Mother or the family you left behind, who would be dead by the time the signal arrives. The information would be out of date by the time it arrives, but at least the general trends of Earth history could be seen by distant travelers. On a star 100 light years from Earth, your Now for Earth would be 100 years delayed, but that would be YOUR reality of what is going on, on Earth. Philosophers can then debate which Now for Earth is meaningful to you. Earth will affect you and your emotions based on what was happening on Earth 100 years ago. The events actually happening 'now' are utterly irrelevant to how you will behave tomorrow.

What is worse is that your technology will steadily become antiquated compared to where Earth currently is in its development unless you have the resources to improve on what technology you currently have and independently keep up with Earth. Earth has a reservoir of millions of engineers and scientists, while you have far fewer. If you embark on a mission using your existing technology, it is likely that a few decades after you start, your spacecraft will be overtaken by a newer-technology, speedier, more efficient, and cheaper design! How will you feel about that, and the years or decades you have perhaps needlessly invested in the voyage so far because you decided to head out into space a bit too soon?

Communication

One of the best science fiction studies of the consequences of long time delays in space travel can be found in R.M. Robinson's *The Seeds of Aril* trilogy. Based on the exploits of the Mizello family, it spans the years from 2149 to 131,000 as explorers are placed in suspended animation using 'magnetocryogenic' technology. Hundreds of one-man ships deriving their thrust by *'extracting electromagnetic energy from the ambient interstellar medium'* made trips between surveyed planetary systems lasting many thousands of years. When travelers were re-animated, contact with local re-supply stations provided the most recent news from Earth and reports from other travelers, that were typically thousands of years out of date. Nevertheless, a Galaxy-spanning human civilization in space emerged from this network of travelers spanning incredible spans of time, long after Earth-bound civilizations had come and gone several times.

Sending and receiving a message even from a distance of 50 light years is no trivial technological task. It is completely dictated by the laws of physics that we have known since ca 1930 for radio communication, and since the 1960s for using optical (laser) methods.

Radio communication

Our corner of the universe is very noisy, with lots of sources of radio-frequency energy such as supernova remnants, star forming regions, pulsars and the general din of charged particles spinning in the galactic magnetic field. It has been known since the 1970s that this background noise has a minimum of intensity between 1 and 10 gigahertz in the microwave region. This is

Interstellar Travel

where virtually all ground-based radio telescopes searching for 'ET' have focused their search.

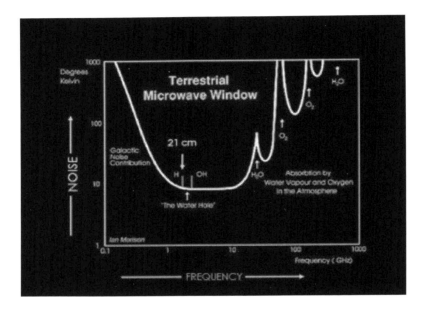

Whenever you point a radio telescope in a specific direction of the sky, it will only pick up signals from a small spot defined by its beam pattern. That means that the signal from our interstellar travelers will be seen against a much larger area occupied by emissions from many other natural sources of radio energy. This figure shows the frequency spread of these natural signals detected by a telescope on the surface of Earth. At these wavelengths, there are no areas of the sky that are free from the unwanted radio noise from the universe.

This image shows what the sky looks like at the 21-cm wavelength of hydrogen at 1.4 GHz. It is completely awash in

Communication

the background noise from interstellar hydrogen, but for most directions in the sky, this is the least amount of background noise that interstellar communication by radio will encounter near the 'water hole'.

Some of the terrestrial noise can be removed if we use a space-based system, but those would be more expensive to build and maintain for only a modest improvement in signal transmission and recovery. The transmission of a signal requires power, focusing and bandwidth, and is also strictly controlled by the familiar inverse-square law.

When a radio 'dish' transmits a signal, most of the power is radiated through the primary beam of the dish, but the rest leaks out in other directions through the secondary and tertiary side-lobes of the dish. The ratio of the power transmitted through the primary to the side lobes is called the antenna gain, so for the highest efficiency and least waste of energy, you want a high-gain antenna. This is why dishes are used because more than 70% of the radiated power goes into the main lobe.

The angular width of the main lobe is given by the formula

Theta (degrees) = 69.9 Lambda/D

Interstellar Travel

$$\text{wavelength} = \frac{\text{speed of light}}{\text{frequency}}$$

$= (3 \times 10^{10} \text{ cm/s}) / (3 \times 10^9 \text{ Hz})$
$= 10$ centimeters.

Since 100 meters = 10000 cm:

Theta $= 69.9 \times (10 \text{ cm}/10000 \text{ cm})$
$= 0.07$ degrees

$1 \text{ ly} = 9.5 \times 10^{12}$ kilometers

$50 \text{ ly} = 50 \times 9.5 \times 10^{12}$ km
$= 4.8 \times 10^{14}$ km

Theta = 0.07 degrees
1 radian $= 360/2\pi$ degrees
$= 57.3$ degrees

$$\frac{0.07 \text{ degrees}}{57.3 \text{ deg}} = \frac{X}{4.8 \times 10^{14}}$$

$X = 5.8 \times 10^{11}$ km $= 0.06$ ly

Operating at a frequency of 3 gigahertz, the wavelength is 10 centimeters, and with a dish diameter of 100 meters (300 feet!), you get a primary beam that subtends an angle of 0.07 degrees on the sky. That is an angle about ¼ the diameter of the full moon!

At a distance of 50 light years, the width of this beam would be about 0.06 light years in diameter! That also means whatever energy you put into transmitting the signal from Earth, 70% of the energy broadcasted by your transmitter will be spread out over a circular area 0.06 light years (580 trillion meters) in diameter! For every watt of energy you transmit from 1 square meter of your dish, only about $0.70/\pi(580 \text{ trillion}/2)^2$ = 2.7×10^{-30} of the original watts is intercepted by 1 square meter of the distant receiver.

If you transmitted a message from Earth to the colonists 50 light years away using a 100,000 watt transmitter and a 100-

Communication

meter dish, it would arrive at its destination as a signal with a strength of $100000 \times 2.7 \times 10^{-30} = 2.7 \times 10^{-25}$ watts/meter^2. If this signal was collected by another 100-meter dish, its strength would be π (100 meters/2)^2 $\times$ $2.7 \times 10^{-25} = 2 \times 10^{-21}$ watts. This signal would be just barely detected against the background noise of the receiver and the various cosmic backgrounds.

So, interstellar communication via radio is doable so long as the travelers have a 100-meter dish and at least a 100,000 watt transmitter. Here is what a 100-meter radio dish looks like:

This is the 100-meter Robert Byrd telescope at Green Bank, West Virginia. It has a mass of 17 million pounds (8500 tons) of machined steel, which is required to steer the telescope and maintain its precise parabolic shape so that it has a well-focused beam. The surface area of the telescope is a 100 by 110 meter

Interstellar Travel

active surface with 2,209 actuators (a small motor used to adjust the position) for the 2,004 surface panels. The panels are made from aluminum to a surface accuracy of better than 0.003 inches. The actuators adjust the panel positions to correct for distortions due to gravity as the telescope moves.

A 100,000 watt transmitter operating at microwave wavelengths is also a feature of radar systems used to track weather systems and military applications. Most advanced weather radar systems now use megawatt transmitters to obtain high time resolution data because the radar rotation rate can be speeded up between refreshes and still get enough signal from distant weather systems to build up a good image and Doppler profiles. These receivers have a mass of a few tons.

In 1963, the USS *Oklahoma City* carried two SPG-49 tracking radars for the Talos missile system. It was a complex collection of electronics in a 19 foot high moving platform that weighed 22 tons. The antenna radiated 3 megawatt of pulsed microwave energy at a frequency of 0.8 GHz. Its working detect range was about 300 kilometers to get a trackable return signal with a 3-meter dish.

Modern megawatt 'X-band' systems can reach 50 megawatts or more because the power is pulsed into separate packets of energy. For example, 1000 joules of electromagnetic energy in a burst that lasts 0.001 seconds equals 1 megawatt of power. The SBX-1 radar complex deployed in 2006 and used for the US Military's ballistic defense program, has an 22-meter diameter dish/array that delivers megawatt pulses. It can detect a baseball from a distance of 4,700 km. The entire system including the

Communication

ship's hull has a mass of about 50,000 tons, but the actual radar itself is probably only 10,000 tons or less.

The 100-meter, 8000-ton telescope with a megawatt transmitter might make good sense on the spaceship, but there would be no way to get it to the ground at the destination. It would remain permanently in space and have to be used there to keep the link to Earth open. A second link could relay the information to the planet using nothing more than a typical 18-inch satellite dish seen on many houses these days.

Laser-optical communication

One problem with the radio system is that the transmitting dish has such a big beam that the power is distributed over a huge area at the destination. The transmitted power from a 100-meter

Interstellar Travel

dish on Earth will be spread out over an area with a diameter of 580 trillion meters at 50 light years.

Laser beams, however, are far more intense and can be greatly focused at their much shorter wavelengths. For example, the Apache Point Observatory Lunar Laser-ranging Operation uses a 3.5-meter telescope to transmit pulses of laser light to the moon and to detect their return pulses. The pulse starts out 3.5 meters in diameter as it exits the telescope and is about 2 km wide at the moon.

The angular resolution of this telescope at 1 micron is Theta=69.9 x (10^{-6} meters/3.5 meters) = 0.00002 degrees. At the distance of the moon (384,000 km) this is a circle with a diameter of about 100 meters, but atmospheric effects on Earth

Communication

broaden this to about 2 kilometers. At a distance of 50 light years, this beam would be about 10^{11} meters across. Centered on our sun, it would cover the entire orbit of Earth! By using larger 30-meter optical telescopes, the beam could in principle be narrowed to smaller than the orbit of Mercury! The challenge is that both Earth and the spacecraft would need such large optical telescopes to concentrate the laser light, and the telescope would have to accurately track a spot 50 light years away to an accuracy of about 4 arcseconds. For a telescope of this size, this is doable using a variety of electronic and mechanical techniques. Some typical 3.5 meter ground-based telescopes have a mass of about 36 tons, most of which is the mounting. By comparison, the 3.5-meter Herschel space telescope weighs 300 kg. It is designed to operate at much longer wavelengths than optical, so its focus is much larger than an optical beam area.

How much laser power do you need in order to be seen at 50 light years? You can buy a commercially available 10 kW 1060-nm-emitting YLS-10000-Y13 fiber laser from *IPG Photonics* (Oxford, MA). These are used for spot-welding! At 50 light years, this energy emitted by a 3.5-meter telescope would be spread out over diameter of 10^{11} meters, so its brightness would be about 10000 watts/π $(10^{11}/2)^2$ = 1.3×10^{-18} watts/meter^2. A similar 3.5-meter telescope would capture about π $(3.5/2)^2$ x 1.3×10^{-18} = 1.3×10^{-17} watts of energy. Now at this wavelength, each photon of light carries 2×10^{-19} joules of energy. Since 1 watt = 1 Joule in 1 second, this works out to be about 70 photons per second!

Of course we don't just get 70 photons every time. The statistical variance on 70 photons is just its square-root or +/- 8

Interstellar Travel

photons. So these 8 photons (roughly 1-sigma), assuming there are no other sources of noise, means that about 68% of the time the received signal will contain between 62 and 78 photons. About 95% of the time (2-sigma) it will contain between 54 and 86 photons, and 99% of the time (3-sigma) you will get between 46 and 94 photons. This is the same kind of statistical variability you get in surveys of people's opinions. For 1000 people the 'margin of error' is just 100%x sqrt(1000)/1000 = 3%. For these kinds of weak signals, the best strategy is to send the same signal many times and average the corresponding data bits. If you average 64 of these 70-photon packets, you beat down the statistical noise to 8 photons/square-root(64) = +/- 1 photon.

This statistical variation is why you need lots of photons to build up a message you can actually read.

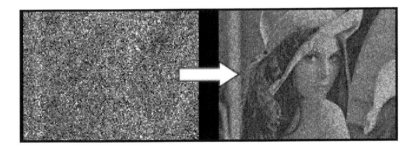

The digital camera image on the left was obtained with about 50 photon/sec, while the one on the right was taken with 2500 photons/sec.

But wait....there's more!

Communication

How do these 70 photons from the distant Earth transmitter compare to the brightness of the sun? The sun is a star with a surface temperature of 5770 kelvins. Every square meter of its surface emits a stream of photons that is strictly defined by a mathematical formula called the black body curve or 'Planck Function'. It is a mathematically precise calculation to determine how many photons each square meter of the solar surface emits at the same wavelength as a 1 micron laser. Actually, the laser emits its radiation over a small range of wavelengths about 0.1 microns wide centered at 1 micron, so we have to collect the solar photons over that span of wavelengths from 0.9 to 1.1 microns.

The answer is that between 0.9 and 1.1 microns, the sun emits 2.2×10^6 watts/meter²/steradian. A full spherical surface has an angular area of 4π steradians (square radian), where one radian = 57.3 degrees. The radiation is emitted out into space so the hemisphere angular area facing space is 2π steradians. The other 2π steradians face the interior of the sun. The light energy flowing out from the surface is $2\pi \times 2.2 \times 10^6$ watts/meter² or just 1.4×10^7 joules/sec/meter².

Photon energy:
$$E = \frac{h \times c}{\text{wavelength}}$$
where
$h = 6.6 \times 10^{-34}$ joules Hertz
$c = 3 \times 10^8$ m/s

For wavelength = 1 micron

$E = (6.6 \times 10^{-34})(3 \times 10^8)/1 \times 10^{-6}$
$= 2 \times 10^{-19}$ joules

The energy carried by a single photon at this wavelength is just 2×10^{-19} joules/photon.

Interstellar Travel

The flow in photons in this band is just $(1.4 \times 10^7$ joules/sec/meter^2) / (2×10^{-19} joules/photon) = 7×10^{25} photons/sec/meter^2.

Now, the entire surface area of the sun is 6×10^{12} km^2 or 6×10^{18} meter^2, and so the total emission of photons out into space is just (7×10^{25} photons/sec/meter^2) x (6×10^{18} meter^2) = 4×10^{44} photons/sec. At the distance of 50 light years, (D=4.8×10^{17} meters) the surface area of the sphere centered on the sun is $4\pi D^2 = 2.9 \times 10^{36}$ meter^2, and so the number of photons crossing this distant surface is (4×10^{44} photons/sec) / (2.9×10^{36} meter^2) = 1.4×10^8 photons/sec/meter^2. Your 3.5-meter telescope that just caught 70 photons each second from the laser, also caught N= π(3.5 meter/2)^2 x (1.4×10^8 photons/sec/meter^2) = 1.3 billion photons each second from the sun! How do we deal with this?

We can either move the transmitter and receiver far away from the sun like beyond the orbit of Jupiter, or we can filter out the sunlight by using a very narrow-band laser and a filter that just admits this narrow range of laser wavelengths. The example assumed a band width of 0.2 microns centered on a wavelength of 1 micron. We can imagine a much narrower band only 1/100 that wide, but we still get 130 million solar photons for every 70 signal photons. This hardly helps enough to matter.

Really, the best solution is to put the transmitter and the receiver as far from their parent stars as possible, that way the entire communication system is protected from sunlight in the first place!

Communication

Now for some really big stuff!

The preamplifiers of the National Ignition Facility are the first step in increasing the energy of laser beams as they make their way toward the target chamber. In 2012, NIF achieved a 500 terawatt shot - 1,000 times more power than the United States uses at any instant in time. Lasers that operate continuously are pretty 'cool' but you can pack a whole lot more energy into a laser beam if you pulse it! Consider the world's most powerful laser, housed in the National Ignition Facility at the Lawrence Livermore National Laboratory in California. It is capable of producing laser blasts with an energy totaling a petawatt (10^{15}, or a quadrillion, watts) and more advanced lasers planned for the next decade will approach an exawatt (10^{18} watts).

Such powerful bursts can only be achieved in short pulses, but for interstellar distances these pulses would be very easy to detect even with bursts lasting only a nanosecond! Modern

Interstellar Travel

photon-counting detectors (photomultipliers) are more than capable of detecting nanosecond flashes.

So, without even knowing whether interstellar travel is technically possible, we have already come up with a communication system that is light-weight and efficient (infrared laser). We know about how big it has to be to reach across 50 light years of space (3.5 meter telescope and a 10000 watt, 1-micron laser) and we also know where we have to put the darn thing to get the cleanest possible signal (outside the orbit of Jupiter) so that we don't compete with the billion-times-brighter sun at the same wavelengths! This system could be as heavy as 30 tons on the ground, but light weight construction in space could bring the mass down to only a few tons.

Once built for space, our communication system would have to stay there, so we will need a way to relay the signal from its distant location to where the colonists will be staying. This is literally off-the-shelf technology today given that we already receive messages from the New Horizon spacecraft out by Pluto! We also know that the same digital signal will have to be sent multiple times and averaged bit by bit to beat down the various noise components.

Things that Break

Very few complex things that humans make last for decades or longer. Including the smallest screws, a typical car has about 30,000 parts. As you well know, cars last perhaps a few years before something goes wrong and you have to bring them to the shop to at least change the oil or fix things that you never even knew existed until the mechanic tells you about them. Almost all of these parts are mechanical and suffer from friction and environment, but there are also a lot of electronic components that can become 'buggy' for unknown reasons.

The Saturn V rocket, fully loaded, weighed 6.2 million pounds and had more than 3 million parts. The Space Shuttle had about 2.5 million parts.

The international space station, built in space for $100 billion, contains over 100 million parts, and various systems have to be replaced every year by ferrying replacements from the ground in shuttle craft. We try to make things very 'reliable' through careful quality control and fabrication techniques. Let's look at the mathematics.

A modern laptop or PC computer has circuit boards that contain literally billions of transistors, diodes, resistors and other nano-scale components. An Intel Core i7 (Ivy Bridge class) CPU has over 1.4 billion transistors on its own. In terms of larger components you can actually see with the human eye, there are just under 100 parts, many of which can be seen on the system 'motherboard' and at least in principle can be repaired, but no

Interstellar Travel

one does that anymore. It is simpler and cheaper to just buy a new computer, except on an interstellar trip!

Suppose you had one part that was manufactured to be 99.9% reliable against failure each year. That means that after 2 years its reliability would be 0.999x0.999 = 0.998 or 99.8%. After 100 years it would be reliable at $(0.999)^{100}$ = 0.905 or 90.5%. This is still not too bad for a 100-year trip and a 1 chance in 10 that it would fail next year. We can reduce this failure rate further by making it 99.999% reliable at the outset, so that after 100 years its reliability is still 99.9% with a 1 chance in 1000 that it will fail.

But what happens if we have two parts with the same reliability of 99.9%? After two years, their probability of still working is $(0.999 \times 0.999)^2$ = $(0.999)^4$ = 0.996 and after 100 years it has dropped to 82%. To bring their combined reliability back up to 0.999, we have to start out with each part having a reliability of failure after first year of 0.999 = $(R^2)^{200}$. If we solve this for R we get 0.999997. So at the start of the mission, each of the two components has R=1 chance in 330,000 of failing each year. Some components like screws and bolts can probably boast that kind of reliability, but what about computer electronics? In this calculation, I assumed the probabilities were independent of each other, but that is a simplification. It is often the case that if one component fails, it takes down some other component so that the probabilities are what statisticians call 'conditional'. In this photo, we see a microATX motherboard with an AMD Athlon 2.10 Ghz processor. Each of the large microprocessor squares contains billion of separate parts far smaller than a human hair!

Things that break

When we consider the millions of components likely to exist in an interstellar spacecraft, failures will be very common because it is impossible to keep up with the demand on making 'perfect' components that never fail after one year. That means crews and computers have to be constantly vigilant for anomalies. But when one is found, it is likely that Murphy's Law will prevail. The part that breaks will be the part that you have no replacement for and cannot repair. This happens all the time in the International Space Station.

The longer the mission proceeds, the more of these irreparable parts will accumulate. Solutions will be jury-rigged (like Apollo 13) until at some point the mission 'fails'. Even if it does

Interstellar Travel

succeed in reaching its destination, you will still not have replacements for your most sophisticated electronics. There are no Intel Corporations to sell you a new motherboard. Basically what has happened is that the distant colonists will steadily lose the advanced technology that was once a part of their daily life as all of the replacements they brought with them fail themselves. There is, however, a solution to this.

Make sure that you have a shot at repairing your systems by forsaking advanced technology in the first place! Go back to the old days of vacuum tubes and thumb-sized components like those you find in grandfathers old radio from before the 1950s. Make sure that the vacuum tubes you use for designing the room-sized computers are all the same kind, or no more than a handful of different types. That way you can bring a boat load of spare parts, and perhaps when you reach your destination you can fabricate them yourselves with a simple repair system you might actually bring with you. All you need is a vacuum system, remote manipulators and spare components found inside cannibalized tubes

Remember, on this journey, there is no real penalty for something being big rather than microminiaturized! If it's big, it might be repairable by you or your children. If it is nanotech, good luck with that. It will take you many generations to re-create a laboratory to repair nanotech components. Even the portable replicator for creating spare parts will eventually break down!

The AN/FSQ-7 Combat Direction Central computer was built between 1963 and 1979 for the U.S. Defense Department. Each

Things that break

of the 24 computers had a mass of 250 tons, used 60,000 vacuum tubes, consumed 3 megawatts of electricity, and had a speed of 75,000 operations per second. Its memory size was 0.3 megabytes. A modern iPhone 6 uses 2 billion transistors, weighs 10 ounces, can perform 800 million operations per second, and has 1 billion bytes of memory. It uses 20 nanometer lithography,

which means the wires and devices are only about a few dozen atoms wide and can be crowded onto a chip only about 1 cm square. You can buy this miracle of technology for a few hundred dollars!

It is enormously tempting to go with the idea of using modern technology rather than vacuum tube technology, but although travelers and colonists have the option of repairing vacuum

Interstellar Travel

tubes, no such option exists for modern technology, so the design of an interstellar trip has a very important choice to make in terms of the technology it will use. Shown below is one of hundreds of vacuum tube modules from an IBM 701 computer developed in the 1950s.

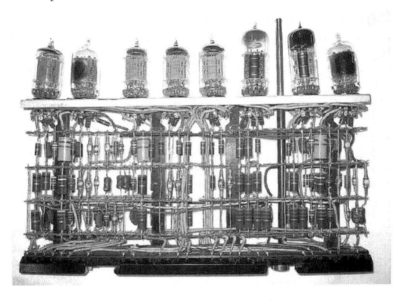

Note, during World War II, vacuum tubes were repaired in the field. The most common way that they failed was that the filament broke. If the ends of the filament were close together, they could be repaired by welding. The repair lasted only a few days to a month, enough time for a replacement to arrive. With modern laser technology, the repairs can be done through the glass. Also, vacuum tubes are sealed by a glass vacuum container, but in space vacuum tubes can be external to the ship without glass enclosures making them even easier to repair. The

Things that break

outside shell of the ship would in principle have many acres of these naked tubes open to space and easy to repair using micromanipulators.

Is this a possible solution to fast but reliable computing in deep space? The 'portable' 450-pound Bendix G-15 computer was introduced in 1956 and used 450 vacuum tubes. It had a 7,800 byte memory drum for storage, and could perform about 2000 operations per second. It was suitable for small-scale computations, and it was accessed by a separate keyboard from which command strings were entered. Programs could be stored on punched paper tape, and read back into the computer memory for execution. The draw back was that, the more complex the program, the less memory there was for actual data storage. But at least in principle, all of its components, vacuum tubes, resistors, capacitors, and switches could be repaired or easily replaced.

Interstellar Travel

We can't know in advance what will break, so we prudently bring with us all of the components that could fail, or we equip all critical systems with multiple levels of redundancy. So, this means instead of one ton of equipment, we have to also drag along additional tons of spare parts to hedge our bets that we can replace any critical component. This is the old tried-and-true approach that has been used for decades when it comes to using complex technology in the field. But what if there were another paradigm? What if we could manufacture replacement parts along the way? At the end of the trip, perhaps less than 1% of the parts we brought along as spares would actually be used, but we don't know which ones in advance, so we have to bring tons of ultimately unused components with us. If we could manufacture the parts, we need to bring a much smaller quantity of raw materials with us!

The big game-changer in is the very recent technology of 3-D printing. All you need is a printer unit and a set of detailed part schematics and you can replicate in a few hours the part you need from raw materials, by what is called additive manufacturing! This is not science fiction, but if it were, it would resemble the *Star Trek* replicator units that create your evening meal, but could also be co-opted to replicate just about anything else you could create a fabrication diagram for.

Things that break

This kind of break-through technology is happening even today.

In April 2015, NASA printed the first copper rocket part; a rocket lining that operates at thousands of degrees and at high pressure. The lining is threaded by hundreds of small channels through which liquid hydrogen flows to keep the 5000 Fahrenheit lining from melting.

Printing organic tissue and some organs like kidneys, liver tissue, bones and similar homogenous organic systems is now being pursued aggressively using the patient's own cells, thereby avoiding rejection issues. This  technology is in its infancy, but in a few decades will be fully mature and will forever change the process of organ transplantation.

Printing microstructures with features a few hundred nanometers in size could be useful for making heart stents, parts for microfluidics chips, and scaffolds for growing cells and tissue. Another important application could be in the electronics industry, where patterning nanoscale features on chips currently involves slow, expensive techniques. 3-D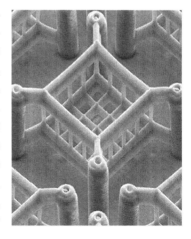

Interstellar Travel

printing would quickly and cheaply yield polymer templates that could be used to make metallic structures.

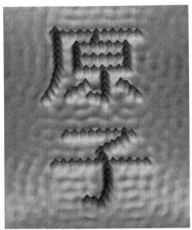
Kanji characters for 'atom'.

Finally, why stop at a nanometer-scale to repair and replace spacecraft nanotech systems. Why not print at the atomic scale? We can already do this with a technique called Scanning Tunneling Microscopy, which can be 'turned around' to actually manipulate individual atoms. Here we see iron atoms on a copper surface forming the

In 2014, the International Space Station's 3-D printer manufactured the first 3-D printed object in space, paving the way to future long-term space expeditions. Testing this on the station is the first step toward creating a working "machine shop" in space. This capability may decrease cost and risk on the station, which will be critical when space explorers venture far from Earth and will create an on-demand supply chain for needed tools and parts. For the printer's final test, in less than a week, a ratchet wrench was designed, approved by safety and other NASA reviewers, and the file was sent to space where the printer made the wrench in four hours.

It's cold in space!

Not surprisingly, for an interstellar trip you will be spending most of your time far from the warming rays of a star, and that is a huge problem.

By the time you get out to the orbit of Pluto, the temperature of a surface exposed to direct sunlight will be only 40 kelvins (-233 C), and at a distance of 8,000 AU (1.2 trillion kilometers or 0.1 light years), the temperature will have fallen to about 3 kelvins. This is the temperature of the fireball radiation left over from the Big Bang and you can get no colder than that. There is, however, one other source of radiant energy that could help you out a bit: Starlight.

Our Milky Way contains over 200 billion stars. If you were to look at the light from the entire galaxy, it would resemble a K-type star with a temperature of about 4,000 kelvins. In interstellar space where you are far from any individual star, the combined light from all these stars adds up to about 8×10^{-8} watts/m^2. This produces about 4 kelvins of heating, which is at least a tad greater than the 3 kelvins of the cosmic fireball radiation, so there you are, roasting at a temperature of 7 kelvins.

So, as you voyage through interstellar space, your spacecraft will be very cold outside, warmed only by diffuse starlight to a temperature barely above the equally-frigid cosmic fireball radiation! Does this present any problems? Unless the starship

Interstellar Travel

design is a sealed 'can' with no external mechanisms, there will be lots of problems getting any mechanical system to operate at near-Absolute Zero. Not even our interplanetary spacecraft have to operate mechanically at essentially liquid-helium temperatures! Here are some records for coldest operating devices and materials.

Temperature		Device or material
-4 F	253 k	LCD displays
-20 F	244	Automobile
-31 F	238	Smartphone
-46 F	230	ISO VG15 hydraulic fluid
-65 F	219	Synthetic hydrocarbon grease
-110 F	194	Silicon greases
-125 F	186	Mars rover
-129 F	184	Vostok, Antarctica
-238 F	123	Silicon transistors
-320 F	78	MOSFETs
-396 F	35	JWST systems

Beyond mechanical issues, how will your spacecraft deal with the ultra-cold conditions of space lasting decades or even centuries? The warm interior of the ship will inevitably leak energy into the cosmos, so to conserve precious energy, you will need to insulate. Any loss of energy over time will not be tolerable because energy is expensive to create...even thermal energy from waste heat has to be recycled. You cannot afford to let it leak away into the universe, which is what the laws of thermodynamics will want to do. Any temperature difference will drive a transport of energy, and the biggest temperature difference is between the outside of the ship at 7 kelvins, and the inside of the ship at room-temperature. The image below

It's cold in space!

shows the heat radiation (infrared emission) from an insulated house. Dark areas indicate where little emitted power is leaking out, while red is where the maximum emission is occurring. The non-black colors represent valuable energy being wasted into space. For homeowners, it represents money while for interstellar travelers it represents bad spacecraft design.

The two fastest forms of losing heat energy are by convection and conduction, which will be important in the inner shell of the spaceship closest to the warmth of the habitation region. Both can be defeated by adding a vacuum shell before the outer surface. The third form of heat transport is by radiation, and this will require more than an efficient vacuum layer in the bulkhead.

Infrared radiation is electromagnetic in nature, and we commonly refer to it as heat radiation. Any surface warmed to room-temperature through conduction or convection will emit this radiation at wavelengths between 2 microns and 15 microns.

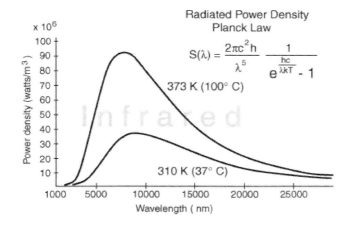

Interstellar Travel

This radiation will travel through the vacuum gap in the bulkhead and reach the outer bulkhead, where it will deliver energy to the outer surface and then leak away into space. To stop this, we have to reflect this radiation back into the ship.

There are many different, commercially available coatings and materials that can reflect thermal infrared radiation. These are used as on glass to prevent leakage through ordinary 'double-paned' house windows. They are also used in fabrics to render humans invisible to infrared cameras. A new material made of cotton coated in silver nanowires, effectively traps heat inside a person's clothing.

The window on the left uses IR-blocking technology to reflect the thermal infrared radiation back into the house, while the window on the right lets it pass into the environment. The ideal interstellar ship would look like the nearly-black window on the left, emitting none of its thermal energy into space so that it can be recycled. Most commercial coatings are even now 80% efficient in reflecting thermal IR radiation.

It's cold in space!

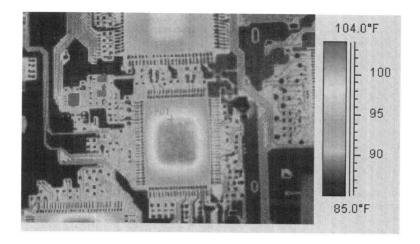

Here is an example of a circuit board. Most components run slightly above room-temperature, but a few draw significant amounts of power such as microprocessor chips, which are warm to the touch and glow at infrared wavelengths. Excessive thermal radiation is probably a sign that your device is too hot and that you need a better design. On an interstellar voyage, all systems must be made energy-efficient to avoid degrading useful energy into unusable thermal energy.

This energy flow diagram for U.S. energy consumption in 2010 is an example of the kind of meticulous energy planning that will go on onboard an interstellar spacecraft. The gray color indicates waste heat produced in the generation process, most from electricity conversion, which is only about 30% efficient. Making the area occupied in the flow diagram by waste heat as small as possible is a desired goal. For the US Energy use in 2010, about 57% was lost as waste heat, mainly in the electrical and transportation sectors.

Interstellar Travel

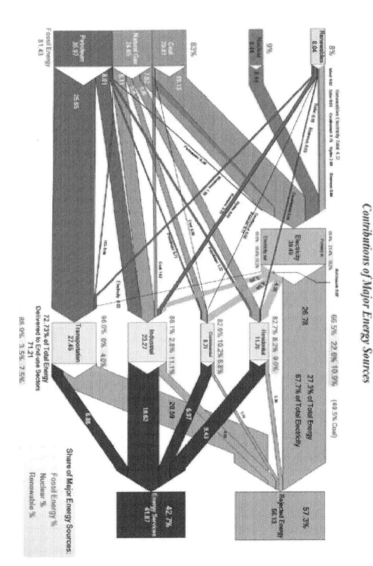

Fuel and Energy

Fuel and energy

No matter where you end up, the same considerations that lead you to a habitable world will lead you to a source of daytime solar energy, which can be recovered either through the high-tech route of using solar panels, or by tapping the photosynthetically-stored energy of the local biosphere plant life. In the latter case, the energy can be recovered directly through chemical combustion, or by fermentation into other flammable compounds (sugars, alcohol, methane). If water is plentiful it can be electrically or chemically dissociated into oxygen and hydrogen gas, which may be cooled to produce 'rocket fuel'.

Depending on the age of the biosphere, there may not have been enough time for significant deposits of fossil organic compounds (coal, oil, methane) to have formed, so it is not reasonable to expect that these will be available. The processing of rock to extract uranium and other energy ores requires a substantial investment of labor, and sophisticated methods for separating the radioisotopes from other elements. This also seems like an unlikely investment of effort at the outset.

Earth has radioactive ores on its surface because of a variety of planetary-scale events, the most important of these is that it has a hot interior and a convecting mantle, which can bring the heavier elements that settled to the core, up to the surface for human access. At the time of formation, these radioactive elements were originally distributed throughout the interior of

Interstellar Travel

Earth, but the subsequent heating eventually partially liquefied the interior. This allowed a process of differentiation to occur in which heavier elements settled to the deep interior. However convection currents brought some of this enriched material up to the base of the lithosphere, where it could be conveyed to the surface through volcanic activity.

One can expect that a distant planet may also have evolved in a similar way if it has a similar mass and size as Earth. In fact, some may have a more vigorous convection cycle and more intense continental drift and volcanism. Others with a more sluggish cycle may have a thicker crust, less volcanism and less vigorous continental drift. The level of volcanic activity might be deduced from the amount of carbon dioxide in the planet's atmosphere.

Geothermal energy – Can use volcanic steam to run a bladed turbine, that rotates a shaft connected to a generator. This only works if your planet has a thin crust, vigorous continental drift and volcanism that is easily accessible.

Wind energy – Wind vanes rotate a shaft that runs a generator. This is readily accessible, but may be erratic and not reliable.

Hydroelectricity – Water mills can grind wheat, or turn a shaft that runs a generator. Planets whose surfaces have running fluids provide the necessary motion. Even the frigid Saturnian moon Titan has a running fluid of liquid methane that can be harnessed!

Fuel and Energy

By far the most reliable source of energy is solar energy. This can be harvested through solar panels, or by bringing your own photosynthesizing organisms that can be turned into fuel. Solar panels are not a technology that can be easily reproduced by colonists until they can construct the proper industrial infrastructure in which to build them locally. They will not have been used on the interstellar voyage because solar panels require far more sunlight than can be found in interstellar space. The raw materials for solar panels are themselves not common. You need pure silicon or germanium as the main substrate.

Although colonists will not have access to high-tech fabrication techniques, they will be able to generate solar electricity as experimenters did at the dawn of this technology on Earth. In 1883 Charles Fritts created the first solar cell by taking a piece of selenium and coating it with a thin layer of gold. It was only 1% efficient but it worked!

Selenium is distributed widely in nature and is found in most rocks and soils at concentrations of about 2.0 ppm. That works out to one gram for every ton of rock. Selenium can be conveniently created as a byproduct of copper refining. Selenium is contained in the constituents of the copper anode during the copper refining process and accumulates on the bottom of the electro-refining tank. These constituents contain roughly 5–25% selenium and 2–10% tellurium.

Even earlier than the Fritts experiments, the French physicist Antoine-César Becquerel in 1839 discovered the Photovoltaic Effect while experimenting with a solid electrode in an electrolyte solution. He observed that a voltage developed when

Interstellar Travel

light fell upon the electrode *"the production of an electric current when two plates of platinum or gold diving in an acid, neutral, or alkaline solution are exposed in an uneven way to solar radiation."*

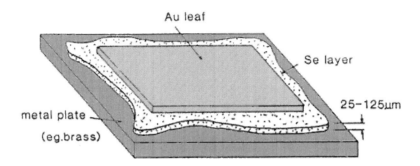

One fundamental problem with all solar-electric systems is that they rely on the photoelectric effect, which is a quantum mechanical phenomenon. A photon with a specific energy causes an electron in an atom to be ejected into what is called the conducting layer of a material. The minimum energy needed is called the band gap. The exact band gap energy depends on the specific quantum properties of the compound. If there are no photons with this critical energy (frequency) then absolutely nothing will happen no matter how bright the light is.

For example, this figure shows the spectrum of our sun (5,770 kelvin), which has its maximum brightness for its temperature at a wavelength of 0.6 microns in what we call the visible spectrum. The blue curve shows the wavelength response of conventional silicon solar cells (c-Si), which have their highest response near 1 micron. A new kind of amorphous-silicon solar cell (a-Si) has a response more nearly in line with how our sun emits light. For interstellar travelers arriving at their destination,

Fuel and Energy

if the star is a cool red dwarf with a temperature of 3,500 k its sunlight peak will be closer to 0.8 microns, so amorphous-silicon cells will not give nearly as much electricity as the conventional-silicon cells. You will need to match your energy technology to the color of the star before you get there, or you may discover you are getting little, or even no, electricity for your effort!

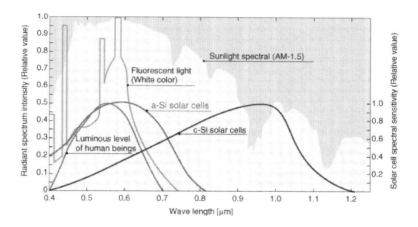

Thermocouple solar electricity – In 1821, Thomas Johann Seebeck discovered that a temperature difference between two dissimilar conductors produces a voltage. Thermoelectric generators can be applied in a variety of applications. The thermal-to-electric energy conversion can be performed using components that require no maintenance, have inherently high reliability, and can be used to construct generators with long service free lifetimes. This makes thermoelectric generators well suited for equipment with low to modest power needs in remote

Interstellar Travel

uninhabited or inaccessible locations such as mountaintops, the vacuum of space, or the deep ocean.

Recently, engineers have created a flat-panel solar-heated electric generator using the Seebeck effect. It achieved a peak efficiency of 4.6% under typical ground-level illumination by the noon-day sun, which means one square meter would generate about 46 watts of electricity.

Free Rocket Fuel?

As your spacecraft travels through the interstellar medium, which is mostly made of free hydrogen ions (protons) and dust grains, there is nothing that says you can't try to collect this material as you travel and perhaps later on use it as 'rocket fuel' or 'reaction mass'

Astronomers have already figured out about how much of this material there is along most directions in the sky out to a few hundred light years. The density of the hydrogen nuclei is between 0.01 and 1 proton per cubic centimeter. We can even create a map of the sky and plot how the density of this gas looks! To figure out how much you can collect is not too difficult. It depends on three things: First, how fast are you traveling? The faster you go the more of the material flows by you for you to collect. Second, how big is your Catcher's Mitt? If you are going to collect it, you have to have something to gather it with. The area of your collector perpendicular to your direction of motion will tell you how much you can collect. If the area is large, you collect a lot. If it's small…well not so much! Finally, it depends on the density of the material in space.

Fuel and Energy

When we put all three of these together we get a formula that looks like this: Number = Area x Density x Speed or N=ADS. You can see that this formula is a good first estimate because when you multiply the units of A, D and S together you get Number = (Square-meters) x (Atoms/cubic-meter) x (meters/sec) which gets you the number of atoms per second that you capture.

$N = A \times D \times S$

$R = 1km/2 = 500$ meters
$A = \pi (500m)^2 = 7.9 \times 10^5$ meters^2
$= 7.9 \times 10^9$ cm^2

$D = 0.1$ atoms/cm^3
$S = 0.1 \times 3 \times 10^8 m/s = 3 \times 10^7$ m/s

Number $= (7.9 \times 10^9)(0.1)(3 \times 10^9)$
$= 2 \times 10^{18}$ atoms/sec

Mass $= 2 \times 10^{18}$ atoms x 1.6×10^{-27} kg
$= 3.8 \times 10^{-9}$ (kg/s)x 3.1×10^7 (s/yr)
$= 0.12$ kg/yr

The amount of mass you capture is just the number of atoms per second x the mass of a proton, which is 1.6×10^{-27} kilograms. For a plausible example, you capture about 2 billion billion atoms per second, which works out to about 0.12 kilograms per year.

If you could eject this mass at 90% C in your engine, it would provide about 2.5×10^{16} joules/year or 800 megawatts of kinetic energy if the energy conversion is 100% efficient. That is nearly 10% of what we need for our 100-ton ship to reach 20%C. Every little bit helps!

Even better, suppose you could fuse these protons to generate energy. This actually doesn't work very well because the energy yield will be miniscule. The sun uses proton-proton fusion at a temperature of 15 million K and actually catalyzes some of the protons through what is called the CNO-cycle. The problem is

Interstellar Travel

that at the low densities and temperatures in the core of the sun, the reaction requires that one of the protons spontaneously turns into a neutron so that the combined fused nucleus has one proton and one neutron forming a deuterium atom. This conversion for a given collision only happens every few million years, but because the sun is so massive, this is enough to allow the reaction to work to stabilize the sun. If it happened faster than this, the sun would have 'burned up' soon after it formed! Well, for the paltry few protons that we collect from the interstellar medium each year (120 grams!), there is virtually nothing we can do with them to generate energy by fusion.

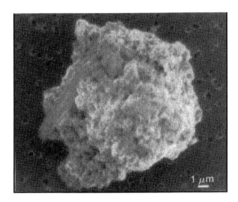

The Ulysses spacecraft found interstellar dust grains with masses between 10^{-20} to 10^{-11} kg, with most near 10^{-16} kg.. These dust grains arrived at a rate of 0.00015 per second across a square meter of Ulysses collecting area.

At 20% of the speed of light, the dust arrival rate will be proportionately higher and about 0.6 dust grains per square meter per second. If we scale this dust arrival rate to a collector 1 kilometer in diameter, with each dust grain having an average mass of 10^{-16} kg, we will accumulate only about 2 grams per year. This is 60 times less than the interstellar hydrogen collected across the same area. So, dust is certainly free mass but not in useful quantities!

Fuel and Energy

The Bussard interstellar ramjet is a popular interstellar ship design based on scooping up interstellar hydrogen gas as fuel. For likely estimates of interstellar gas and dust density, it will probably not have access to enough free fuel to make it work, especially for the high-tonnage ships likely to be needed, unless collecting areas over 4 km across are used for 100-ton spacecraft designs. These large collecting areas also increase the ship's friction with the solar wind, and may even prevent the ship from leaving the solar system in this way.

Antimatter !

Interstellar Travel

Antimatter was first proposed in the early-1930s by physicist Paul Adrian Maurice Dirac as a result of developing a quantum theory for matter that also obeyed Einstein's special relativity. In 1932, Carl Anderson experimentally confirmed the existence of the anti-electron (called the positron) in cosmic rays, and this opened up a whole new search for the antimatter versions of all other elementary particles. The next pairs discovered were the anti-proton and anti-neutron, which were manufactured artificially in 1955 by Emilio Segre and Owen Chamberlin using the new Bevatron at U.C. Berkeley. This accomplishment won them the Nobel Prize for physics.

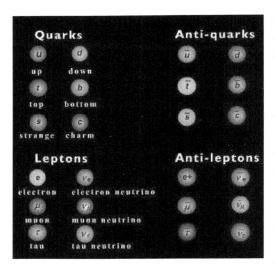

Antiparticles are identical in mass to normal particles, except that all of their quantum numbers (e.g. charge and spin) have been reversed. For example, the electron has a negative charge but the positron has a positive charge. The electron neutrino has no charge, but it does carry a quantity known as the Lepton Number, which is conserved in all interactions. Electron neutrinos have a Lepton Number of +1 and antineutrinos have a Lepton Number of -1. For quarks, they

Fuel and Energy

also carry a color charge, which is what the strong nuclear force connects to the way that the electromagnetic force connects to the - and + charge in electrons and protons. Quarks come in three colors: red, blue and green, and have charges of -1/3, and +2/3, but it is only the electric charge that defines the antiparticle. So an Up quark has a charge of +2/3 but an anti-Up has a charge of -2/3. For example, the neutron consists of two Down and one Up quarks. The Down quark has an electric charge of +1/3 and the Up quark has a +2/3 charge. The anti-neutron, even though it has zero electric charge, consists of two anti-Down and one anti-Up quark. These have charges of -2/3 and +1/3.

Finally, we have the particles that cause the three basic forces in nature, and some of these have their own antiparticles. The electromagnetic force is transmitted by the photon, which is its own antiparticle. The strong force is transmitted by eight gluon particles. Each gluon has a hybrid color like red/blue, red/green, and the gluon antiparticle is anti-red/anti-blue, anti-red/anti-green and so on. The weak force is transmitted by three particles: W^+, W^- and Z^0. The Ws are antiparticles to each other, and the Z^0 is its own antiparticle.

The thing about antiparticles is that, when they are brought into contact with their normal matter partners, the quantum numbers (charge and spin) cancel-out so that the total becomes zero. This happens explosively. Because energy has to be conserved, it appears in the form of two photons which carry no charge (-1 and +1 = 0), but because they carry a spin of 1 unit, the photons spin in opposite directions to cancel out this quantum number. The electron has a spin number of +1/2 and the positron has a

Interstellar Travel

spin number of -1/2 so the net spin equals zero, but a photon has a spin number of +1, so we have to emit a second photon with a spin of -1 so that the net is also zero. So, when an electron (+1/2) and a positron (1/2) annihilate, two photons of opposite spin are produced that each carry the mass of an electron, which in energy units equals 511,000 electron-Volts/c^2. Physicists usually drop the c^2 because it is understood to be a part of the mass unit, and so if you want to look for signs of electron annihilation, you just look for its signature energy of 511 keV. This photon energy makes the photon an x-ray.

One important detail about our universe is that the matter that came out of the Big Bang was normal matter. We do not live in a universe where there is an equal amount of anti-matter. This was all sorted out before the first second after the Big Bang. For every 10 billion antimatter particles, there was about 10 billion and one normal matter particles. When they combined, the 10 billion antimatter and matter pairs annihilated to produce 2 billion photons. These became part of the cosmic fireball radiation of the Big Bang. The left-over one particle of matter added up to become all the normal matter in our lop-sided, matter-dominated, universe.

Today, the universe creates antimatter momentarily in the cores of stars where the thermonuclear reactions bring them into existence, and in certain kinds of collisions of cosmic rays with matter in interstellar space. But this is nothing like the production during the Big Bang. There are no large quantities of solid antimatter in our universe. They could never have survived contact with normal matter in the closing moments after the Big Bang.

Fuel and Energy

And, oh yes, we can artificially create antimatter in our laboratories using very expensive equipment. But we create it literally one particle at a time. Here is a photograph of a particle entering from the lower-left and colliding with an atom, which stops its forward momentum. The collision liberates enough energy to produce an electron-positron pair. Because their charges are opposite, they spiral in opposite directions in the magnetic field in this detector.

Currently, the annual production of antiprotons at the Antiproton Decelerator facility of CERN was several picograms at a cost of $20 million. To get just one kilogram at that rate would take about 1,000 grams/(10^{-12} grams/yr) or about 1000 trillion years.

There's no point in even estimating the cost!

So the bottom line is that there are no minable 'chunks' of antimatter that we can find in our local universe, and we can't make the stuff in large enough quantities to matter in our ship design, so once again, Nature has given us an energy source with one hand, but taken it away with the other. There is nothing in the laws of physics that prevents us from using antimatter the

Interstellar Travel

way some engineers would like. It's just that we do not live in a universe that has this resource available to us today.

Without being too concerned about the fuel cost or availability, here is one imaginary design for an antimatter rocket. Antiproton annihilation reactions produce charged and uncharged pions, in addition to neutrinos and gamma rays. The charged pions are channeled by a magnetic nozzle, producing thrust. This type of antimatter rocket is called a pion rocket or a beamed core configuration. It is not perfectly efficient; energy is lost as the rest mass of the charged (22.3%) and uncharged pions (14.38%), lost as the kinetic energy of the uncharged pions (which can't be deflected for thrust), and lost as neutrinos and gamma rays.

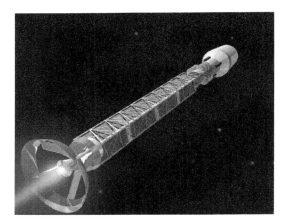

So, antimatter holds a lot of promise as an energy source, but there is no current scalable technology that would produce it in the quantities we need for interstellar travel. But there is another problem. Creating antimatter is one thing, but if it comes into contact with normal matter, like the air in your room you can't

Fuel and Energy

see, it instantly vanishes in a burst of energy! You have to contain and isolate the antimatter while you are creating it. The only two ways we know to do that are to store it in a perfect vacuum, or suspend it in a powerful magnetic bottle.

Magnetic bottles only work if the antimatter is charged, and as you collect more and more of it, the pressure of the plasma increases until the magnetic field starts leaking. Controlled fusion research has been trying to confine dense, charged plasma in magnetic fields for decades, but the confinement is usually violated by complex plasma instabilities and leakage effects.

To appreciate the scale of this problem, pictured here is the Tandem Mirror Experiment built by the Lawrence Livermore Radiation Laboratory in 1979. This machine was built to understand high density plasma confinement inside magnetic fields. It was a room-sized device that had a mass of 350 tons, used a 125 kiloGauss magnetic field, and confined a few trillion ions per cubic centimeter - far less than a microgram of matter!

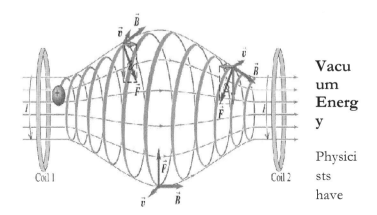

Vacuum Energy

Physicists have

Interstellar Travel

known for decades that space is not really empty at all. Thanks to Heisenberg's Uncertainty Principle, which is the foundation of all quantum theory, you can never determine exactly how much energy is in a given volume of space during an observation that lasts a finite amount of time. It is sometimes called the Quantum Embezzlement Principle. You can notice one dollar missing for weeks, but to lose a thousand dollars it would only take you a few hours to notice that. Within this principle, the vacuum is no longer empty at all because energy is equivalent to mass ($E=mc^2$). Within the quantum vacuum, pairs of particles and their antiparticles literally flash into existence, annihilate, and vanish in a burst of photons that themselves cannot be detected. A mathematical model of what this looks like is shown below.

All normal matter like electrons moves within this constantly shifting patina of complex virtual interactions. Sometimes these virtual particles interact fleetingly with normal particles and the

Fuel and Energy

minute push or pull causes a particle to erratically change its direction. Here's a diagram that shows two electrons approaching each other from the top of the diagram, they exchange a photon and are bumped away from each other. The

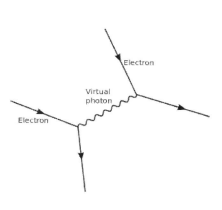

photon of light drawn as a wiggly line is not a real photon like the kinds that come out of a flashlight. They are virtual photons hidden in empty space. In this case, we only draw one of trillions upon trillions of these virtual photons that actually are interacting with the two electrons. Taken together, this swarming cloud of virtual photons is what we call the electric field of the electron. You know that like charges repel each other, and this diagram describes that result according to the modern language of quantum mechanics. All forces are caused by the exchanges of particles that 'mediate' the force and that normally are present as unobservable virtual particles.

Sometimes during a violent collision between ordinary particles, we can give these virtual particles enough energy to become real particles with positive-energy. At other times we can do this by colliding a matter particle with its antimatter twin. They annihilate into a dense fireball of pure energy, which can then impart enough energy to 'empty space' to bring into existence a real pair of particles and antiparticles. The pair has to have no net charge or spin, because both of these quantities were also

Interstellar Travel

annihilated when the original particles came together. Physicists like to say that the resulting real particles have to have the same quantum numbers as the vacuum. So long as this rule is obeyed, you can spawn just about any kind of pair of particle and antiparticles from the vacuum. Here's what happens when you collide two protons at the Large Hadron Collider in CERN at an energy of 7 TeV (7 trillion electron-Volts or 1×10^{-6} joules per particle)!

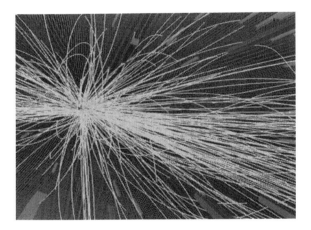

Over 100 new particles were ejected from the raw energy of the head-on collision, including the two original protons. These particles, when summed together will have no net charge or spin like the vacuum from which they originated. There will be equal amounts of particles and antiparticles too. But now things can get very interesting. The virtual particles in the vacuum act like a form of wave energy at many different wavelengths. If you put a pair of conducting plates into space, separated by a few microns, some of the wavelengths will not penetrate this small volume of shielded space. That means the energy of the vacuum between

Fuel and Energy

these plats is slightly less than outside the plates, and this causes a miniscule pressure and force that tries to push the plates together. So, even though we cannot detect these virtual particles directly thanks to the Heisenberg's Uncertainty Principle, we can still measure their effects on matter. This Casimir Effect was first proposed by physicists Hendrik Casimir and Dirk Polder in 1947, and actually measured 1997 by physicist S. Lamoreaux. At a separation of a few hundred times an atomic diameter, the pressure can be as high as air pressure at sea level! Here is a diagram showing the experimentally-measured force strength (in trillionths of a Newton) when plates are separated by less than 500 nanometers (0.5 microns).

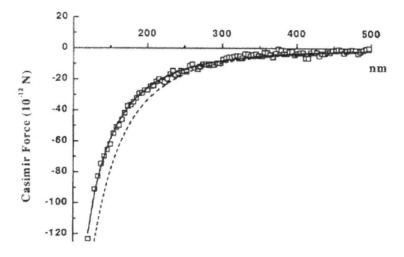

The theoretical formula shown in the solid line is $F = -(\pi^2/240)(Shc/a^4)$ where S is the area of the plate, h is Planck's constant, c is the speed of light and a is the plate separation in centimeters.

Interstellar Travel

For example, two plates each with an area of $S=0.25$ micron^2 separated by a=0.2 microns would feel a repulsion of $F=-2 \times 10^{-8}$ dynes or 20×10^{-12} Newtons.

$$F = - \frac{\pi^2 \, S \, h \, c}{240 \, a^4}$$

$S = 0.25$ micron$^2 = 2.5 \times 10^{-9}$ cm^2
$A = 200$ nm $= 200 \times 10^{-7}$ cm

With:
$h = 1.05 \times 10^{-27}$ $c = 3 \times 10^{10}$ cm/s

$F = - [(3.14)^2 \times (2.5 \times 10^{-9}) \times (1.05 \times 10^{-27}) \times (3 \times 10^{10})] / [240 \times (2 \times 10^{-5})^4]$

$F = -2 \times 10^{-8}$ dynes of force

Because it represents a force, that means in principal it can be used to do work and therefore generate energy. This is called zero-point-energy (ZPE), and there is considerable controversy as to whether it can be extracted from the vacuum in quantities large enough to be useful. Some experiments in nanotechnology are underway to try to harness this force to make frictionless surfaces, but physicists believe ZPE is useless in anything like industrial quantities.

The main problem is that, when you extract this energy the vacuum has to settle into an even lower energy state than 'zero energy', and that violates not only thermodynamics but quantum theory. There is no lower energy state than the quantum vacuum WITH its myriad of virtual particles.

Harnessing ZPE would be a very neat trick for interstellar travel, because then you would no longer need to bring any fuel with you. You just extract it out of 'empty space' as you go. There have been many science fiction designs for how to do this that

Fuel and Energy

are very entertaining and inspiring. But like the fabled Perpetual Motion Machine, and for the same reasons, this curious effect in the atomic-scale laboratory will not work for us at the scales we need.

This is probably a good thing because one might imagine the unintended environmental effect of extracting energy from the vacuum. The vacuum might reach a tipping point and implode.

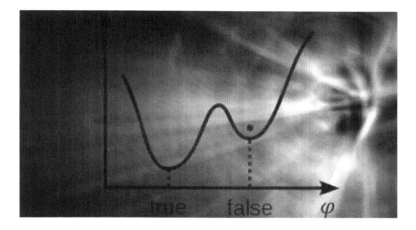

That is NOT a good thing!

In cosmology we deal with the energy of the vacuum state. It is as much a part of our real world as sunlight. This energy is determined by an energy diagram that looks like the one shown above. In a 'true vacuum' state, the energy of space is at its minimum, and space expands by doubling in separation as time doubles in duration. This is called a linear progression, or in our case the normal Hubble expansion for the universe. When the universe was young, it was temporarily in a 'false vacuum' state

Interstellar Travel

called Inflation. As it transitioned over to the lower, true vacuum state, as time durations doubled, space separations increased by factors of two or more in an exponential progression. Turning this around, if you detect an exponential expansion of space, you are in a false vacuum state. Currently, our universe is exponentially expanding due to the presence of Dark Energy.

Now, back to our story!

If you were to extract zero-point energy from the vacuum, you would create a region of space with slightly less pressure than its surroundings, and it would also have a lower energy than the current vacuum energy. The result would be an implosion of space. Whether this would escalate to something even more catastrophic because a 'tipping point' or threshold had been crossed is unknown.

Down and Back

.

From the interstellar spacecraft you can do all the remote sensing of your new planet that you want. You can map the entire surface to Google Earth-quality resolution. This image of animals at a waterhole in Africa was obtained from Google Earth. The resolution is about 70 centimeters, and uses the DigitalGlobe Quickbird imaging satellite system. These satellites operate at about 600 kilometers above the ground, and their imagers cover a surface area of 680,000 km^2 every day. The optical system uses a 1-meter primary mirror, and the satellite has a mass of three tons and consumes 3 kilowatts of solar power.

You can even send down robotic probes like Curiosity on Mars, to sample the geology and whatever biosphere there is. But at some point you have to decide that this is the place you want to

Interstellar Travel

settle, or at least visit first-hand before the long return to Earth. This presents us with a grave problem.

As countless astronauts can now tell us, making the trip to the surface of a planet is not that hard. It has now been accomplished hundreds of times using Apollo-like capsules, and even sophisticated Space Shuttles. The problem we now face is that to get back up to our interstellar spacecraft after the expedition is over requires a large rocket, which we have to bring with us on the trip here! Scaling from the 8-person capacity of the Space Shuttle, not everyone on the spaceship will be able to travel on such a round-trip, and perhaps only a handful will form the expeditionary force. This also assumes that they will be physically fit enough to make the switch to operating under a high-gravity environment.

We might imagine disassembling the rocket in space and sending its parts down by parachute for reassembly, but to reassemble a rocket and stack its components requires a launch complex like the Kennedy Space Center with large cranes and lots of people to do the work. This is hardly the job for the handful of people you just sent down!

If we were able to miraculously transport a one-shot rocket down to the planet's surface so that our explorers could return to the spacecraft, would this even work?

We have all watched breath-taking Space Shuttle launches, where solid fuel rockets are used to supplement the Shuttle main engine to reach an altitude of about 100 miles, then the main

Down and Back

fuel tank supplies the Shuttle Engines with fuel until it is ejected after about nine minutes. After that, the onboard fuel tank

supplies the Space Shuttle with fuel to take it into orbit. During the Apollo Program, we used a Saturn V rocket and a second

Interstellar Travel

stage to place the third stage and Apollo Command Module into a trajectory to the moon. This approach works because of the Rocket Equation,

V = Exhaust speed x Ln(Rocket mass/final payload mass)

> For example, suppose you needed to reach an escape speed of 10 km/sec and your exhaust speed was 1 km/sec. You have a 1-ton payload. From the rocket equation:
> 10km/s = 1 km/sLn(X/1tons)
> Solving for X we get
> e^{10}=X/1ton
> So we need a one-stage rocket with 22,000 tons of propellant. Multi-stage rockets are far more efficient than this because they eject excess rocket mass.

which precisely describes how much fuel and mass you need to leave the surface of a planet using a one-stage system. The key ingredient that describes the ratio of the payload mass to the rocket-plus-fuel mass is the surface gravity of the planet. The more fuel you need to overcome the higher gravity, the smaller the payload mass can be.

For the Saturn V multi-stage rocket, the payload mass (Command Capsule plus Service Modules) was 120 tons to Low Earth Orbit. The total mass of the fueled rocket was 2970 tons, for a payload-to-mass ratio to LEO of 4%. That's for a planet with 1 G at the surface. For a planet with 2Gs at the surface, and the same diameter as Earth, the escape speed is now $(2)^{1/2}$ = 1.4 times greater, because escape speed V = $(2GM/r)^{1/2}$. This means that the payload-to-rocket mass ratio has to be reduced by a factor of ¼ so that now we are only bringing 30 tons to orbit for a launch vehicle as big as the Saturn V. We are now down to a payload mass that is 2% of the entire launch vehicle.

Down and Back

For a planet with 4Gs at the surface, another factor of two, we are now only bringing ¼ of 30 tons or 7 tons into LEO with the Saturn V and we are now down to a payload mass of about 1%. Of course actual engineering calculations and higher-performance (thrust to weight) engines can improve this considerably, but the point is that at some value for the planet's surface gravity, we will not be able to leave once we arrived on the surface, and this applies to planets with higher than Earth gravity where a return flight becomes increasingly likely as the colonists health under high-G begins to decline. Unless we find a planet with almost exactly 1-G at the surface, we are in for some very grim prospects for returning to the orbiting interstellar ship in a single multi-stage rocket, let alone the dozens of rockets needed to transport an entire colony.

Another possibility is to create a Space Shuttle-like transport system that carries its own fuel for the round trip, and both lands and takes-off like a jet plane. This does not sound as crazy as it seems. Even today there are plans for designing such hypersonic shuttles that take off from the ground and achieve the Mach-20 speeds needed to leave the planet.

The 'scramjet engine' is the key enabling technology for sustained hypersonic flight. Propelled by this special type of air-breathing jet engine, a high-performance hypersonic craft might even be able to fly into orbit. Because neither scramjets nor ramjets can operate efficiently when they are traveling below Mach 2 or 3, a third type of propulsion (perhaps turbojet or rocket) is required for takeoff. So-called rocket-based combined-cycle engines, which could be used in a space vehicle, rely on a rocket that is integrated within the scramjet combustor to

Interstellar Travel

provide thrust from takeoff through subsonic, low-supersonic and then ramjet speeds. Ramjet operation is then followed by scramjet propulsion to at least Mach 10 or 12, after which the rocket is utilized again to supplement the scramjet thrust. Above Mach 18, the rocket by itself propels the vehicle into orbit and enables it to maneuver in space. NASA is currently testing several variations of such a system

Here we see an artist's conception of the X-43A Hypersonic Experimental Vehicle, or "Hyper-X" in flight. The X-43A was developed to test a dual-mode ramjet/scramjet propulsion system at speeds from Mach 7 up to Mach 10.

Down and Back

The only drawback to these systems is that you need a well-designed runway BEFORE you can make the first landing, which is a chicken-and-the-egg situation.

The unavoidable aspect of leaving a planet or returning to it is that it takes raw energy to make the trip. For Earth, we need to give a rocket and payload a speed of about 10 km/sec to reach orbit. Because $E = 1/2mV^2$, this works out to a minimum of 50 million Joules per kilogram. It can be higher than this depending on the efficiency of the launch vehicle. We also have to lose this much energy to return to the ground, but this can be done by dissipating the energy through atmospheric friction and heating of the external 'heat shield'. For more massive planets, the energy increases as the square of the escape speed and is proportional to the mass of the planet, so a planet twice the mass of Earth would require twice the energy per kilogram. At some point we have to confront the problem of fuel mass, engine mass and thrust, where we need low-mass fuel, a low-mass engine technology, and a very high thrust limited by the speed of light. Currently only chemical rockets deliver high enough thrusts for their mass to get us off a planet.

Interstellar Travel

How much will it cost?

What do really big engineering projects look like, and cost? How about projects that require the development of new technologies as they go? It's easy to pick out the really big projects and estimate how much they cost. Engineers call them 'megaprojects' because they cost upwards of $1 billion to complete and garner a huge amount of public attention.

Close to the top of this list is the Apollo program to put astronauts on the moon at a cost of $203 billion (2015). Next we have the Space Shuttle program at $204 billion (2015) followed by the International Space Station, which cost by some estimates $150 billion (2010).

In terms of civil engineering, China is planning to spend $370 billion between 2015-2025 on atomic energy, and we have the

Interstellar Travel

construction of the Khazar Islands in Azerbaijan for $100 billion, which will house 1 million people. The Winter Olympics of 2014 cost Russia about $50 billion (2015). If we did the Manhattan Project again it would cost $30 billion (2015).

Looking back at some historical projects, by some estimates it would cost $5 billion to build the Great Pyramid of Khufu, and an unimaginable $300 billion to re-create the Palace of Versailles in France. In terms of infrastructure, by 2014 we had over 1200 operating satellites in space, representing a worth of about $200 billion in hardware placed there over a span of 20 years. But how much would it cost to re-create the entire North American roadway system which took 100 years to build piece-by-piece? The Congressional Budget Office says that in 2007 alone, the public sector invested $146 billion to build, operate and maintain its highway system

How about re-building the entire computer industry and technological base starting with the work by Alan Turing in 1940? In 2015, the cloud computing market is estimated to be about $121 billion by itself. The size of the world's software industry in 2013 alone was estimated to be $1.4 trillion.

In 2013, the Credit Suisse Global Wealth Report says that the personal wealth of the world is $241 trillion. This wealth is not distributed equally among nations, as we all know. North America and Europe account for just over half of all the wealth ($152 trillion) with Asia-Pacific ($48 trillion) and China ($22 trillion) accounting for most of the rest. It is expected that by 2018 global personal wealth will increase by 40% to $334

How much will it cost?

trillion. Surprisingly, only $1.5 trillion of this wealth is spent on military investments.

By comparison to these great projects, our investment on other goods and services are far from trivial. For example, according to the USDA estimate in 2006, Americans spend $154 billion every year on alcohol. In 2015, it is estimated by the American Pet Products Association that we will spend $60 billion for pets, food and medical treatment. In 2013 Americans spent $13 billion on bottled water, and worldwide we spend nearly $100 billion. In 2015, Apple debuted its new 'Watch' and an estimated 40 million people are expected to buy it for a total of $20 billion.

Clearly, we have a huge amount of money in the world economy, although like the value of your mortgaged house, much of it is not in 'liquid' form. But enough is through our annual salaries that we can afford enormously extravagant purchases, and through poor decisions both personal and political, silently condone a tremendous amount of waste. For example, rather than maintain our civic infrastructure through annual and appropriate taxation, a tax-averse society waits for things to break, bridges to collapse and potholes to damage our cars before we make the necessary investments, which are then substantially more expensive.

So what about space travel? The Space Shuttle combined with the International Space Station, which it was designed to build, had a combined cost of about $354 billion, employed more than 300,000 people in high-tech jobs, took 30 years to build and maintain, and will last until about 2025 before being

Interstellar Travel

decommissioned. It did not take us anywhere except Low Earth Orbit, but allowed us to experiment with a 6-man crew living in

space for up to 6-month shifts. The next step to Mars has been estimated at $100 billion, and would require 20 years of developing the right technology. But estimates as high as $230 billion have also been put on the table. If we also factor in additional Mars missions following the Apollo multi-visit strategy, this would raise the total cost to $1.5 trillion.

Compare this with the US Gross Domestic Product for 2014 of about $18 trillion, and you have some sense of the relative costs. Of course the Mars investment of $1.5 trillion would be over 10 years or so, so that we are actually comparing an annual Mars cost of $150 billion to an annual GDP of $18 trillion, which is only 0.8% of the GDP. Would a Mars trip and a handful of preliminary colonization activities be worth 0.8% of the US GDP? As a percentage of the global wealth of perhaps $400 trillion, it represents an even more miniscule investment. Nevertheless, this only gets us to Mars, which in round numbers is barely 100 million kilometers from Earth on a 'bad day', using technology we largely have available to us today. What happens

How much will it cost?

when we dare look to the stars and estimate those travel and colonization costs as best we can?

In 1968, physicist Freeman Dyson discussed the cost of such a program in his paper titled *'Interstellar Transport'* and concluded that the cost could be as little as $100 billion which, adjusted for inflation, is about $650 billion today.

In 2012, Andreas Hein, an engineer with the Icarus Interstellar Project, wrote a detailed paper that estimated the cost for an unmanned, 50-year journey to Barnard's Star using "fusion drive" technology. It consisted of 50,000 tons of fuel and 500 tons of scientific equipment. Top speed: 12 percent of the speed of light. The price tag for this venture was about $100 trillion. This sounds like a lot, but his argument was that when spaceflight is seen as a percentage of the US, GDP of about 0.3% and for the world GDP of 0.08%, he estimated that such a mission would be at that level of our GDP by the 24th century. He also estimated that a 'budget Daedalus' mission to Barnard's Star similar to Freeman Dyson's design could cost about $65 billion.

A subsequent estimate for an interstellar mission by Robert Zubrin, an aerospace engineer and co-founder of the Mars Society also estimated that it could cost $125 trillion to build a ship capable of traveling 10 percent of the speed of light and supporting a few score human voyagers. With an average doubling of global wealth every 15 year, by 2200 the world is 1000 times richer and even expensive interstellar voyages may not seem so economically daunting.

Interstellar Travel

The Daedalus starship design compared to a Saturn V.

"At its zenith, the Apollo program employed 400,000 people and needed the support of more than 20,000 industrial firms and universities. The enterprise of constructing a starship would very likely be abundantly more challenging and quite possibly be 100 to 1000 times more demanding than the Apollo moon program. Thus, an admittedly simple, yet instructive linear extrapolation indicates that construction of an interstellar spacecraft might employ as many as 400 million people – more than the current population of the United States. What seems immediately apparent is that an interstellar vehicle would be a massive source of employment thus representing a huge public service." (Icarusinterstellar.com)

In addition to cost, we have to confront what kind of social system would have to exist to make such a project feasible? Technologically, we have many examples of trillion-dollar construction projects that have been undertaken over decades,

How much will it cost?

such as our commercial and military satellite systems, and our maritime commerce system. What all these big projects have in common is that they provided very useful products and services while they were under construction. We did not have to wait until that last mile of roadway was installed before we could use the entire network.

It is easy to sell a major engineering project when citizens being 'taxed' for it can get some benefit from the program along the way. The initial years of the space program were problematical until military uses were found for it and then commercialization began so that unclassified resources like TV broadcasting and the Global Positioning System could be sold to the public. So long as the point of the effort was just putting a few lucky astronauts into orbit, or snapping a few scientific pictures of the surface of Mars, the public considered it to be a waste of time and money. Luckily space programs are very inexpensive compared to other elements of the federal budget!

For interstellar travel, it will be difficult to market it as an exciting social service or means of developing interesting commercial products. Most of these will have to do with travel and colonization of the solar system, and these are not seen as important goals of Earth society in the face of so many other pressing issues, the least of which is climate change. If you thought selling the US space program has been hard, and that's just to maintain a budget that is 0.8% of the annual US Gross Domestic Product, wait until you have to sell interstellar travel!
The pyramids of Egypt were a piece-of-cake as a social labor program compared to the deferred-gratification of an interstellar spacecraft construction.

Interstellar Travel

According to Marc Millis, former head of NASA's Breakthrough Propulsion Physics Project and founder of the Tau Zero Foundation, interstellar travel may be impossible based on energy costs alone. For a trip to Alpha Centauri, the energy needed to accelerate a 500-person ship to 10% the speed of light is about 10 exajoule (10^{19} joules). That is all the energy consumed by humans on Earth in one year. Even without accounting for fuel, the 500-passenger ship wouldn't be able to launch until around 2200 at the earliest, and the Alpha Centauri probe won't be ready until around 2500.

As impossible as interstellar travel may seem, there are still good reasons to consider it a reasonable technological goal. The return on investment in the US Space Program has been estimated by economists to have been $6 for every $1 expended in the space program. This resulted in technological spin-offs that revolutionized not only consumer technology but health care and other valuable economic activities, not even considering the huge number of new 'high tech' careers that were created for millions of people that never existed before the space age.

The Defense Advanced Research Projects Agency (DARPA), recently created the *100-Year Starship Project* as a one-year, $500,000 to identify the key technologies and approaches to making interstellar travel a reality. They even have a website, an former Space Shuttle astronaut will lead the program. Their goal is

'*We exist to make the capability of human travel beyond our solar system a reality within the next 100 years. We unreservedly dedicate ourselves to*

How much will it cost?

identifying and pushing the radical leaps in knowledge and technology needed to achieve interstellar flight, while pioneering and transforming breakthrough applications that enhance the quality of life for all on Earth. We actively seek to include the broadest swath of people and human experience in understanding, shaping and implementing this global aspiration."

Who can honestly say how the innovative solutions they find may eventually impact our society, whether or not we even take the next step and actually make the journey!

Interstellar Travel

The Miracle Cure: Stasis?

Many of the problems we encounter with human crews traveling great distances, such as psychological problems and resource consumption, can be made far less severe is we transport people from place to place, unconscious.

'Sleeper Ships' as they are called in the science fiction world are a common main-stay for certain kinds of SF writing in which light-speed travel can't yet be attained, or where long voyages are required to get from place to place. For example, the *Alien* series of stories have sleep chambers for the crew, for multi-year journeys.

The benefits of traveling in stasis are enormous. The concept of crew boredom is non-existent, as are many of the other psychological problems that attend cramming many people into a small, monotonous volume of space. Also, if you want to shield the crew, you only have to shield the volume in which their bodies are stored, not the entire volume of the spacecraft. You would also not have to pressurize the entire interior of the spacecraft.

Hibernation

One form of stasis is hibernation. This is a type of reduced metabolism or torpor caused by hypothermia. A 10 degree drop in body temperature reduces metabolic rate by 50 to 70 percent. Astronauts would need some sort of activity to avoid muscle loss during the trip, but even back home, animals as large as

Interstellar Travel

bears suffer very little muscle loss during their hibernation. Perhaps reviving the astronauts periodically so they can 'stretch their legs' would be enough to ward off muscle loss over longer periods of time.

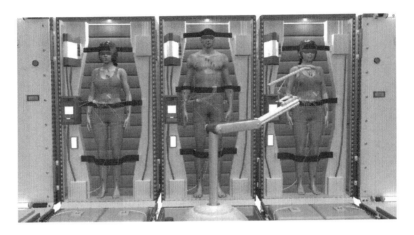

Rather than just an idea in science fiction, NASA is actually looking into hibernation for the 200+ day journeys to Mars. The crew would be kept alive through intravenous feeding. Electrical impulses would stimulate the muscles of the astronauts while hibernating, ensuring that when they woke from their slumber they were not significantly weakened.

A system called *RhinoChill* used in hospitals to induce torpor reduces body temperature by about one degree per hour. It does this by supplying coolant through the nose, and chilling the part of the brain that regulates metabolism. At a temperature between 89°F and 93°F, the crew would enter stasis. To get out of this state, the supply of coolant is stopped so that the patient's body can return to its normal basal temperature. After

The Miracle Cure: Stasis?

being in this state of clinically-induced torpor for a few weeks the crew could be awakened in shifts so that someone could keep an eye on the ships systems and the other hibernating crew. It is estimated by NASA that this could cut the weight requirements of such a mission from 400 tons to about 220 tons.

The good news is that, once you take the body to the minimum stasis temperature, there seems to be no additional benefit to going lower in temperature than about 10° F below the normal body temperature. As for duration, the longest times reported in medical journals are about a week. In one study, a maximum duration of 16.4 days was reported. The most frequent side effects were electrolyte disorders (77%), pneumonia (52%), thrombocytopenia (47%), and septic shock syndrome (40%).

Artificial Coma.

A barbiturate-induced coma, or barb coma, is a temporary coma brought on by a controlled dose of a barbiturate drug, usually pentobarbital or thiopental. Most of the time, this is only used in emergencies where severe brain injury has occurred. The longest induced comas are rarely over six months

The treatment is not without side effects and risks. Infections and poor immune system response is one of the most common; however, some patients report experiencing vivid nightmares and hallucinations that doctors believe are the brain's effort to make sense of the sounds around in its environment.

Interstellar Travel

The problem with all these methods is that they do not really stop the human metabolism, nor do they stop the metaboli of the organisms inside the human body. For every cell in the human body that is 'yours' there are 10 viruses and bacteria. These almost literally have a mind of their own, and that poses a severe problem for long-duration hibernation. It is not out of the question that, while you are in stasis, the other microbes in your body are continuing their millennia-old battles for dominance. There are bacteria and viruses that can go dormant under the right conditions, but there are also others that may find a sluggish human metabolism no problem at all, especially if it causes competing bacteria to go dormant and give up their ecological niches. They may find that munching on human organs and other tissues, like the now-famous *Streptococcus pyogenes,* is just the thing to keep them alive while you are in stasis. Upon awakening, there may not be much left of you except a large mound of organic soil.

Also, these methods slow down metabolism but do not stop it, so that means that your own cells will continue to grow, though more sluggishly with the reduced oxygen and nourishment intake. That poses a problem for individuals with microscopic tumors or cancer cells. Oncologists think that everyone has a variety of cancer and pre-cancerous cells within themselves all the time, but that our immune systems are effective in finding them and removing them. But hibernation may also make the immune system much more sluggish, allowing cancers we could not detect to gain the upper hand. This is especially problematic in a radiation-rich environment like space. Humans are not the only species that has been thwarted in this way during hibernation.

The Miracle Cure: Stasis?

According to a BBC News interview, *"As their body temperature drops, hibernators also remove the lymphocytes (white blood cells) from their blood and store them in the lymph nodes. And within 90 minutes of awakening, these reappear. This damping down of the immune system prevents a general inflammation in the body during rewarming – the very thing that would cause humans and other non-hibernators to suffer kidney damage. However, it's a risky strategy, leaving animals unable to mount an immune defense while hibernating. The fungus responsible for white-nose syndrome, currently wiping out bat colonies in the USA, takes advantage of this vulnerability, infecting the bats while they are dormant. In response, the bats frequently exit hibernation and rewarm to fight off the pathogen – the high-energy cost of these interruptions ultimately killing them."*

Interstellar Travel

Artificial Intelligence and Virtual Reality

The entire discussion so far has been about transporting flesh-and-blood humans to the star as the only goal of space exploration. It is the 'given' and the uncontested assumption behind virtually every science fiction story ever written in the interstellar genera. In some sense, it reflects the Manifest Destiny of the 17th and 18th century outlook on who we are as humans and where our destiny ultimately lies. It is also the grandchild of the idea that humans have literally limitless possibilities as a species. It began when the first human ancestors left Africa 100,000 years ago and walked themselves across Europe, Asia and the Americas, and then paddled their way to Australia and Polynesia. Most humans, however, are content to stay home.

At the time Christopher Columbus ventured to the New World, the human population consisted of 375 million people, and the mission itself used three ships based on 100-year old technology to support 87 crew members on a 3-month voyage across the Atlantic. His second voyage involved 17 ships and 1,000 men. The vast majority of humanity neither knew about this venture or cared, and the cost was completely born by wealthy nobility, King Ferdinand and Queen Isabella of Spain, and a number of investors, including the town of Palos at a cost in modern dollars of about $40 million. There is an excellent article on this: Look for Satava, "Columbus's First Voyage: Profit or Loss From a Historical Accountant's Perspective" in *The Journal of Applied Business Research* Vol 23, Number 4, 2007.

Interstellar Travel

Of course we all know of an earlier exploration by the Viking Leif Erikson around 1000 AD, with one boat and a crew of a few dozen, which landed in at L'Anse aux Meadows in Newfoundland. This was a low-budget voyage that probably cost no more than a million dollars in modern currency, but hey, who's counting Viking currency? Again, the population of the world was about 250 million, and most of these folks had no idea about exploration.

Contrast this with the Apollo landings on the moon. To get there from the vacuum tube technology of the 1940s and V2 rockets, we had to create within a few decades entirely new technologies, rocket motors and miniaturized low-mass computers based on transistors and integrated circuitry. The 21 astronauts who made these landings and the 12 who actually visited the lunar surface for a few days cost an estimated $1.6 billion (2010 dollars) for the Mercury program, $7.5 billion, and the Apollo program cost $110 billion for a grand total of about $119 billion (2010). This was born entirely by the American taxpayer. When we add to this the cost of developing the integrated circuitry that ran the computers, we can easily double this cost. The prevailing mood of the taxpayer was negative, even though this largely politically-driven program was our main argument against the believed backwater progress of communism in the USSR. The entire Space Program was generally seen, not as a scientific activity by explorers but as a shrewd political activity in the Cold War race for world popularity of democracy against the evils of communism and the Soviet threat of world domination expressed at the United Nations by Premier Khrushchev a decade earlier.

Artificial Intelligence and Virtual Reality

Even today, the popularity of space exploration is extremely low. According to extensive studies of surveys by Roger Launius at the National Air and Space Museum, *"Consistently throughout the 1960s a majority of Americans did not believe Apollo was worth the cost, with the one exception to this a poll taken at the time of the Apollo 11 lunar landing in July 1969. And consistently throughout the decade 45–60 percent of Americans believed that the government was spending too much on space, indicative of a lack of commitment to the spaceflight agenda… For example, while Americans may not know much about the space program, they have a largely favorably opinion of it—over 70 percent say they have a favorable impression… A human Mars mission also has never enjoyed much support from the American public. Consistently, more people polled have opposed the mission than supported it. With that lukewarm support the nation's elected leaders will certainly not proceed down this policy path unless something else—probably some crisis—requires it… It seems obvious that most Americans have little conception of the amount of funding available to NASA. At a fundamental level, all federal programs face this challenge as Americans are notoriously uninformed about how much and what the federal government spends on its programs. As a result there is a general lack of understanding that NASA has less than one percent of the Federal budget each year, and that its share of the budget has been shrinking since the early-1990s. Most Americans seem to believe that NASA has a lot of money, much more than it annually receives. Turning around those false perceptions of funding is perhaps the most serious challenge facing those who wish to gain greater public support for space exploration."*

Contrast this with other recent studies that paint a more optimistic picture. According to the Pew Research Centers 2015 survey, *"NASA continues to be very popular among the public, with four times as many Americans holding a favorable view of the space agency as*

Interstellar Travel

unfavorable (68% vs. 17%). In contrast with many other departments and agencies of the federal government, Republicans and Democrats generally have the same positive view. NASA rated at the top of a list of eight government agencies along with the Centers for Disease Control and Prevention in a Pew Research Center survey last month."

What does all this have to do with how we choose to explore space and venture to the stars? The likely cost of interstellar travel will have to be borne by the entire world population. For a $100 trillion mission supported by 10 billion humans, that works out to $2,000 per person over a 50-year development period. Possibly less. Only an overwhelming world support for such a venture will allow the 'majority' to coerce the 'minority' to be taxed at such a rate to support a human expedition to the nearest likely candidate system. Following the Apollo model, we can expect that huge improvements in technology will appear from this venture that will have major commercial application. For one, the likely rocket technology will revolutionize interplanetary travel, making a trip to Pluto in a few days seem likely. But returning to the assumption that humans will make the first interstellar voyages, we open up new possibilities if we relax this assumption.

As we have seen, conscious humans are a major liability and health risk. As a species, we were never evolved to meet any of the challenges of interstellar travel, whether we deal with the collapse of our immune systems in a sterile ship environment, or the likely-fatal boredom of the physical and social environment. But we already have embarked on an entirely new set of technologies that may revolutionize our exploration, if we are only willing to think less parochially about what constitutes

Artificial Intelligence and Virtual Reality

human exploration. Does interstellar travel always mean that we have to send humans?

The most promising technologies for interstellar travel are virtual reality and artificial intelligence. The former makes exploration a truly communal and inclusive activity for billions of people left behind to foot the bills, not just the lucky few that get to make the trip. The latter development of Artificial Intelligence (AI) removes all of the risks of human exploration and allows robotic exploration to take over for domains beyond the solar system. Because these technologies are of interest to humans for other purposes, we are once again in the model of the Apollo program and huge technology spin-offs that can be immediately commercialized and brought in to the lucrative entertainment world for the average human. The modern cellular phone is the direct descendant of Apollo technology, and is now used by literally billions of humans at a cost of many hundreds of dollars per unit, not including 'data plans'.

Artificial Intelligence

In May 1997 something amazing happened. For decades it was thought that one of the quintessential attributes of human intelligence was its ability to look into the future, see possibilities for action, and to plan accordingly in a current action. The game of chess became the tall pole activity that would be the intellectual stand-in for human superiority. Then in 1997, an IBM computer called Big Blue defeated the World Grand Master Garry Kasparov 3 ½ to 2 ½. This computer could evaluate 200 million chess positions every second, and could

Interstellar Travel

carry out 11 billion floating point operations per second (11 GigaFLOPS). In a November 2006 match between a modern chess-playing program called Deep Fritz and world chess grandmaster Vladimir Kramnik, the program ran on a personal computer containing two Intel Core 2 Duo CPUs, capable of evaluating only 8 million positions per second, but searching to a depth of 18 moves into the future. The computer won 4 to 2 by its last match on December 5. Armed with sophisticated algorithms and a massive memory of historical games and gambits, it is now nearly impossible for any human to win chess games against a silicon-based opponent!

Another memory-based performance was won by a computer called Watson on February 15, 2011. This time the game was the highly-watched *Jeopardy!* TV game show with Alex Trebek moderating, and the competing players being Ken Jennings and Brad Rutter. Watson, also created by IBM, was a supercomputer that used 2,880 POWER7 processor cores and had 16 terabytes of RAM. It could process 500 gigabytes per second and performed at a speed of 80 teraFLOPS, some 7,000 times faster than Big Blue in 1997. It cost an estimated $3 million, and it responded to natural-language questions.

The memory capacity of the human brain has been estimated at between one and ten terabytes, with a most likely value of 3 terabytes. Consumer hard drives are already available at this size. It is predicted than by 2030, micro-SD cards with capacities equal to 20,000 human brains will be routinely available!

According to Dr Kurzweil, Director of Engineering at Google, "*The paradigm shift rate is now doubling every decade, so the next half*

Artificial Intelligence and Virtual Reality

century will see 32 times more technical progress than the last half century...Three-dimensional, molecular computing will provide the hardware for human-level 'strong artificial intelligence' by the 2020s. The more important software insights will be gained in part from the reverse engineering of the human brain, a process well under way. Already, two dozen regions of the human brain have been modelled and simulated,"

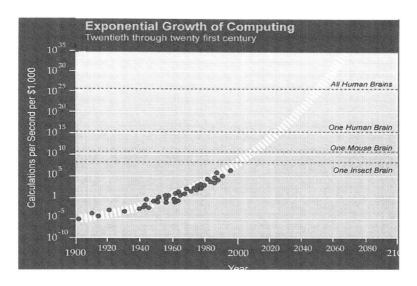

When seen against this backdrop, interstellar travel with AI systems carrying the human banner of progress may be an entirely reasonable and eminently-workable solution if you can keep the total mass for the entire ship plus AI system down to a few hundred tons or less.

Interstellar Travel

Virtual Reality

Virtual Reality (VR), is an immersive multimedia or computer-simulated life that replicates an environment, which simulates physical presence in places in the real world or imagined worlds. Virtual reality can recreate sensory experiences including virtual taste, sight, smell, sound, and touch. I am not talking simple computer gaming, here. VR plays an important role in combat training for the military. It allows recruits to train under a controlled environment, which uses a Head-mounted display (HMD - see photo), data suits, data glove, and VR weapons to train for combat.

You can experience a simple form of VR with data manipulation by using *Google Earth* and using the icon to switch to street view. Imagine using this feature to explore Mars using satellite imagery of the same caliber as Google! This is not as far-fetched as it seems. *Google Mars* and *Google Moon* are already available. Fully immersive experiences are literally just around the corner.

According to R. Terrile at the Jet Propulsion Laboratory, "Once immersive and interactive data sets become available and

Artificial Intelligence and Virtual Reality

commonplace there will be increased pressure on NASA to provide similar experiences from future space missions. The step-wise increase in value of higher fidelity public education products like full color panoramas, stereo views and time lapse movies has been demonstrated as they became ingrained in our culture. Just as live television transmissions from astronauts on the lunar surface four decades ago created an irrevocable expectation for public engagement, so too will immersive telepresence.

The key ingredients are that the scene has to have the resolution of the human retina, and the software has to smoothly keep up with changes in head and body position and direction of stare. Both of these place very high demands on data storage and processing speed. But once they are met, the brain is unable to literally tell the difference between the computer reality generated by telepresence, and normal perception. As Terrile notes *"Immersive telepresence offers the motivation for sending astronauts to the vicinity of Mars (short of landing on the surface) because it allows them to interact and explore the surface with dexterous robotic proxies (like Robonaut 2) without suffering the speed of light latency of operating from Earth. In this way human directed Mars surface exploration can occur earlier and more economically by combining with robotic science missions. It further enables all of Earth-bound humanity to take the journey as well and sense the thrill of walking on a alien world. Three-dimension surface models with real physical properties and used with avatars further allow the simulation of surface deformation. Thereby, the public can virtually explore, dig holes, move rocks and leave footprints on Mars."*

Only this kind of shared exploration experience as a proxy for physically being there makes any sense in the long-run. The

Interstellar Travel

main problem is that familiarity invites boredom once the novelty has worn off unless as any computer gamer will tell you, there is an element of adventure involved. Because the solar system is a finite environment with only a few dozen surfaces that you can physically walk on with robotic proxies, any human will quickly run out of fun things to do. Of course, this will not be the case for the Citizen Scientist, or the pre-college student. Will a stroll on Ganymede take the place of a Friday-night card game with friends as a commonplace cultural activity the way massive online computer gaming battles have become?

It only took a few years from the commercialization of radio in the 1920s, to the habit of millions of families crowded around the family radio listening to their favorite show. This was replaced in the 1950s and 60s with a similar family TV Night. Given the speed with which technology has invaded the Family Room in the year 2015, it is a sure-thing that family journeys off-Earth will become equally commonplace and desired entertainment by the end of this century.

We are a long way from the holodecks of Star Trek-fame, but for now we have the beginnings of immersive VR that lets us physically walk through unimaginable scenery using treadmills or simply cavernous halls. Will climbing walls be added to the mix so that explorers can free-climb canyon walls on Mercury and through their exertion, completely feel the experiences as authentic down to the flexure of a single finger muscle searching for a firm purchase?

Artificial Intelligence and Virtual Reality

These advancements in the VR technology and AI defy no physical law, and they are the direction that entertainment technology is currently advancing. Only the scenery will change.

Just send your brain!

By 2030, most technologists expect we will have arrived at The Singularity. At this time, computer memory technology will have advanced to the point that the entire human brain could be stored in a reasonably compact computer memory technology. In other words, since your brain is a set of synapses or switches, these could be reproduced in a silicon technology so that at least on paper, your brain could be downloaded into silicon. Instead of a human making the trip to the stars, their entire brain could be read-out and telemetered across interstellar space, then an exact copy of their brain would be built at the destination. As seen from the standpoint of the traveler, the trip would have seemed instantaneous. The process, of course would be completely destructive of their former biological bodies and now the person would be some form of disembodied data base (memory) or program (mind) operating in a remote computer system.

According to some estimates, if it takes 3×10^{14} bits to represent a human brain, shipping this data across 300 light years with current technology takes about $4,500. When you add in the radio transmission technology the cost rises a bit more to $50,000 per person per trip. It is hard to imagine how such a system would work, especially since the destination must already have the equipment to reconstitute the radio signal into a silicon-based personality. Somehow, the technology to perform

Interstellar Travel

this operation must already have traveled across space to the destination, which takes us back to the issue of interstellar travel that we started with. We have not actually saved any money in the first step of the process; we have only made subsequent trips

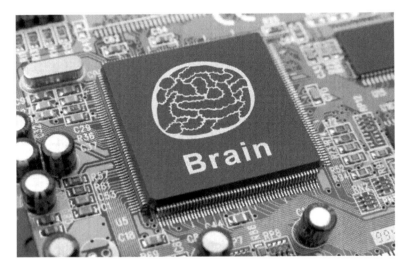

by humans cheaper and have avoided the astronomical expenses of additional interstellar spacecraft. This 'neuromorphic chip', for example, is part of a rapidly evolving technology to reproduce neural activity for selected cerebral functions on a chip. In this instance, the device works with an artificial retina to recognize moving patterns and images.

There is also another issue. Would this new memory storage system also include a complete body with a sensory interface to the brain? Humans are more than a collection of static memories. There must also be a 'program' operating to turn memories coded in synapses into a functioning personality.

Interstellar Rocket Technology

Let's assume that we have solved all the scientific and technical issues about selecting the best destination, and how we will survive once we get there. Of course the BIG QUESTION is, how will we actually make the trip in a reasonable amount of time? Here are some very basic assumptions that we cannot work around no matter what particular technology we invent to make the actual voyage.

Fifty light years: First, the nearest destination where we would expect to find an Earth-sized planet in its habitable zone, orbiting a star that does not produce regular superflares, and which might have a workable atmosphere, though not an ideal one, is <u>probably</u> no closer than 50 light years. Alpha Centauri is eliminated, as are the stars closer than 15 light years because they are, even now, known not to have suitable planets.

One hundred tons: Second, we cannot know how many travelers will make the trip in a conscious, active state. Given that hibernation is an option, we still need enough support to have a stable trip environment for at least a few people working in shifts. As a reasonable, round number, let's say that the habitable, cargo-carrying part of the spacecraft has a mass of 100 tons. This will not be enough to have a lander capable of making a return trip and the fuel to support it, by the way.

Twenty percent the speed of light. Let's assume this will be the top speed. It is twice the 10% C that many estimates assume, but because the stars we want to reach are more than twice as

Interstellar Travel

far away as the ones most studies consider, doubling the speed seems like a reasonable compromise. Besides, if we have the technology to get to 10% C, then finding ways to double that speed are less of a challenge than it might seem!

Because we want to get to our maximum speed as soon as possible, coast for a time, and then reverse our speed back to near-zero so we can orbit the planet, this implies a very specific way in which we are going to increase and decrease the kinetic energy of our spacecraft, and the rate at which we will need to consume that energy to reach the desired speed. At this point, we are talking elementary physics and simple math. Because the top speed is far below the speed of light, we do not need to use relativity to get to answers that are good to better than 10% in our estimates.

> Kinetic energy, $E = \frac{1}{2} mV^2$.
>
> Mass, m, of the spacecraft is 100 tons
> $m = 10^5$ kg,
>
> Top speed is 20%C or 60,000 km/sec
> $V = 60,000,000$ m/sec
>
> $E = 0.5 \times (10^5 \text{ kg}) \times (6 \times 10^7 \text{ m/s})^2$
>
> $= 1.8 \times 10^{20}$ joules.
>
> Power = Energy/time so if
> $E = 1.8 \times 10^{20}$ joules and
> $T = 1$ year $= 3.1 \times 10^7$ seconds, then
> $P = 1.8 \times 10^{20}$ joules/3.1×10^7 seconds
> $P = 5.8 \times 10^{12}$ joules/second
>
> $P = 5.8$ trillion watts

The kinetic energy of a 100 ton spacecraft traveling at 20%C is 1.8×10^{20} joules. That is equal to the energy released by 40,000 megatons of TNT or 2000, 20 megaton atom bombs! If we assume that it takes a full year to boost to that speed, then the rate at which we have to provide power to increase the kinetic

Interstellar Rocket Technology

energy is 5.8 trillion watts. The yearly average energy consumption of the entire United States is 0.5 trillion watts.

Another way to look at this problem is to estimate the average acceleration you need to get to 20% C in one year. That works out to a = speed change/time = (60,000,000 m/s) / (3.1x10^7 sec) = 1.9 m/sec^2 or about 20% the surface gravity of Earth. The acceleration of gravity on the moon is 1.6 m/sec^2 so the ship provides a moon-like experience!

$D = \frac{1}{2} a T^2$

$A = 1.9$ m/sec^2
$T = 1$ year $= 3.1$x10^7 sec

$D = 0.5 \times 1.9 \times (3.1 \times 10^7)^2$
$D = 9.1 \times 10^{14}$ meters

1 light year $= 9.5 \times 10^{15}$ m
So $D = 0.1$ light years.

To fine-tune this calculation, as we accelerate from 0 to 20%C in one year, we will travel a distance equal to 0.1 light years! Near the destination we have to turn the spacecraft around and decelerate for the same amount of time, which adds another 0.1 light years. Since our trip is 50 light years, we only spend about 50-0.2 = 49.8 light years coasting at a speed of 20%c, which will take T = 49.8 ly/.2c = 249 years to traverse, and so adding up the times, the trip takes 249 + 2 = 251 years. If we add relativity to this, our coasting ships time will only be 249*(1-0.2^2)$^{1/2}$ = 244 years for a total of 246 years. That's a full 5 years less than what our Earth-bound observers see.

The problem with this energy calculation is that we have not allowed for the loss of fuel mass out of the 100 ton spacecraft as we boost the speed to 20% c. How do we do that unless we know beforehand what the technology is for the rocket engines?

Interstellar Travel

Only this will tell us what the mass of the engine plus fuel will have to be. Is there any way to figure this out without knowing what the technology might be?

The basic idea behind any rocket is that it throws mass out at a high speed to create the momentum to accelerate forward in the opposite direction.

Charged Reaction Mass

Ion thrusters with the Dawn spacecraft eject xenon ions at speeds between 15 and 35 km/s with a 7 kilowatt engine design. NASA's High Power Electric Propulsion (HiPEP) systems operate at 50 kilowatts, a propellant throughput of 100 kg/kilowatt and produce 0.7 Newtons of thrust, with ion ejection speeds of 100 km/s.

Although these ion thrusters have masses of a few hundred kilograms and consume less than 100 kilowatts of electricity, they are only the tip of the iceberg if you relax the mass and weight restrictions. All of these designs are electrostatic accelerators in which ions are accelerated between two or more grids that have a potential difference of several thousand volts.

Physicists design accelerators that boost particles to more than 99.9% the speed of light, a whopping 300,000 km/s, but they use radio-frequency klystrons in which microwave energy is transferred to the ion in successive stages along a linear waveguide to boost their speed. Small examples of these 'linear accelerators' are used in hospitals to accelerate electrons to energies of 10 MeV. Physicists, however, have devices that

Interstellar Rocket Technology

accelerate heavy ions. The point is that technology exists today to accelerate ions to speeds of 90% c or more for protons and 2 MeV per nucleon (RHIC in 200 MeV accelerator) for gold atoms and up to 9 GeV per nucleon (RHIC).

New accelerator technology is being developed to create table-top accelerators that achieve 1 GeV per meter using charged particles riding on laser beams. The technology is called wake field acceleration. In 2014, physicists at the Lawrence Livermore Laboratory in California used a petawatt laser and succeeded in accelerating electrons to an energy of 4 billion electron volte (4 GeV) in a 9-centimeter-long chamber. The electrons were accelerated by 'riding' the electric field set up by the laser. The laser is guided through a straw-like chamber containing a plasma. The laser also creates electromagnetic waves in the plasma, and electrons become trapped in these waves and are accelerated. It is hoped that this technology will allow physicists to eventually create accelerators like the Large Hadron Collider at CERN, but instead of a ring 27 km in circumference, you will only need an accelerator a few hundred meters long to reach the same energies.

So, there does not seem to be anything to stand in the way of accelerating electrons and even protons and heavy ions to a sizable fraction of the speed of light. Today's methods will eventually be miniaturized and improved even more. Given many promising technologies available today, accelerating protons, electrons and nuclei to nearly the speed of light poses no significant problem.

Interstellar Travel

For relativistic speeds, the amount of kinetic energy you have is given by $E = mc^2 (1/\sqrt{1-(v/c)^2} - 1)$ so for $v = 0.99c$ and m = mass of a proton = 1.6×10^{-27} kg, we have $E = 6mc^2 = 8.6 \times 10^{-10}$ Joules/proton. This also equals an energy of 5.4 GeV with an additional 0.94 GeV for the rest mass of the proton, for a total energy of 6.3 GeV per proton. This is the energy that physicists refer to when they speak about 'atom smashers' like the new CERN, Large Hadron Collider.

Proton energy $E = 8.6 \times 10^{-10}$ joules/proton.

Power = 5.8 trillion joules/sec

Ejection rate:
$P/E = 5.8 \times 10^{12} / 8.6 \times 10^{-10}$
Rate = 6.7×10^{21} protons/sec

Mass rate:
Proton mass = 1.6×10^{-27} kg
Mass rate = $6.7 \times 10^{21} \times 1.6 \times 10^{-27}$
= 1.1×10^{-5} kg/sec.

Current:
1 ampere = 6.2×10^{18} charges/s
Proton current = $6.7 \times 10^{21} / 6.2 \times 10^{18}$ = 1080 Amps.

To accelerate a 100-ton spacecraft to 20%c in 1 year (lunar gravity acceleration) requires 5.8 trillion watts, so if our accelerator produces 8.6×10^{-10} joules/proton, the rate at which we eject protons from the engine has to be about or 6.7×10^{21} protons/sec. That works out to about 0.01 grams/sec or 1100 Amperes! Over the course of a full year, the fuel (protons) you will need is about 310 kilograms to accelerate to 20%c and 310 kilograms to decelerate back to zero, so for the round trip (500-year flight) we need about 0.6 tons. So, the amount of fuel we need is insignificant compared to the 100 tons of spacecraft mass. But we do have to take into account the total mass of the technology we actually need to make the engine physically work.

Interstellar Rocket Technology

What is the mass of an accelerator capable of boosting 310 kilograms of protons to 99% c over the course of a year, and then repeating this miracle at the end of the journey when we arrive?

Accelerators require magnets, klystrons, control electronics and a power source to run the whole shebang. We can't know what these parts will look like in 100 or 200 years from now, but we can at least begin the process by considering what these numbers look like for current technology! Current ion engines have masses of a few dozen kilograms, but they work at such low power that they produce only a slow stream of ions at 100 km/sec or less. On the other hand, a research-grade accelerator that can push protons to 99%c requires a rating of about 6 GeV. In 1954, UC Berkeley physicists built the 'Bevatron' to accelerate 10,000 protons at a time to just this energy. It used a magnet that had a mass of about 10,000 tons and it looked like this:

This is also a far cry from the nearly 7×10^{21} protons/sec we need to supply thrust. But in 60 years we can do a lot better!
The Francis H. Burr Proton Therapy Center at Mass Gen in Boston has a 270 MeV proton accelerator, is 4 meters in diameter and weighs 220 tons using 3-Tesla bending magnets. At this energy, the protons are traveling at about 80%c. They consume about 300 kW of power. A compact medical synchrotron design with a diameter of 4 meters could have 300 MeV (speed = 80%C) with up to 2×10^{11} protons/sec in a 1 cm² beam.

Interstellar Travel

Although even with medical proton accelerators we can get to nearly the speed we want, the number of protons, 2×10^{11} protons/sec, is far lower than the 7×10^{21} protons/sec we need. But medical accelerators are a bad model for comparison because their beams are limited by the dose delivered to the human tissue without incinerating the patient. Research accelerators like the Linac 2 at CERN used for the Large Hadron Collider produce beams at 300 milliAmperes, which is 1.9×10^{18} protons/sec! For our spacecraft we need a beam current of about 1100 Amperes!

So, at least for now, the requirement that we need an engine that produces 1100 Amperes of 6 GeV protons and has a mass significantly under 100 tons, seems like an insurmountable technological challenge. But in 200 or 300 years, who knows? These kinds of systems will not be built unless there is a pressing need for them, and at least interstellar travel is a plausible high-tech need!

> Kinetic energy $E = \frac{1}{2} mV^2$
>
> $V = 5000$ km/sec
> $M = 0.2$ kg
>
> 1 apple $= 0.5 \times 0.2 \times (5 \times 10^6)^2$
> $= 2.5 \times 10^{12}$ joules/apple
>
> Rate $= 5.8 \times 10^{12}$ joules/sec / 2.5×10^{12} joules/apple)
> $= 2.3$ apples/sec.

Uncharged Reaction Mass

Chemical rockets typically eject matter at speeds of a few thousand meters per second. For example, the Space Shuttle exhaust speed was about 3 km/s. This is not going to work for us.

Interstellar Rocket Technology

The reason we turned to accelerating 0.3 tons of charged protons during the acceleration and deceleration phases of the trip is because at least in principle it is easy to accelerate charged particles to nearly the speed of light. But what can we do with uncharged matter? The 5.8 trillion watts of kinetic energy we need, when rendered into ejecting apple-sized (0.2 kg) balls of matter out of the engine requires each apple have a speed of about 5,000 km/sec and we need a steady stream of about two apples every second for 2 years of speed-up and slow-down! We can use other matter streams and they all result in the same acceleration power. For example, at an ejection flow of 10 grams/sec you will need a speed of 35,000 km/sec. The problem is that the slower the speed, the more fuel mass you will need over one year of acceleration. At 10 grams/sec you will need 310 tons of mass per year, and at the slower speed of 5,000 km/sec you will need over 15,000 tons! These masses are so large compared to the 100-ton mass of the spacecraft that we would have to completely re-calculate the acceleration rate and the new fuel requirements.

So, it is very hard to use neutral matter as the reaction mass unless you can reach exhaust speeds near the speed of light, which we can only easily achieve for charged matter. One proposal that has been discussed is to create a series of small nuclear explosions and use the impacting gas on the back of the ship to accelerate the ship forward. These methods also require large quantities of mass exceeding many thousands of tons, and the acceleration is jerky rather than continuous. Although this is not a problem for unmanned journeys, it is a problem for conscious ship crews. One of the original 'Orion' designs by Freeman Dyson required 400,000 tons of deuterium pellet fuel

Interstellar Travel

to achieve a top speed of 10,000 km/sec, an acceleration of 1G and a travel time to Alpha Centauri of about 130 years one way. You would need the same amount of fuel for the return trip.

This picture shows NASA's reduced-size 6,000 ton version of the small-sized interplanetary Project Orion craft with a mass up to 8 million tons. The depiction shows the craft about 4

milliseconds after the explosion of a one-kiloton, nuclear propellant charge. The glow around the base of the craft is the ultraviolet-glowing plasma from the explosive recompressing against the pusher plate.

There have been many references to this nuclear pulse technology in science fiction. *Footfall* by Larry Niven and Jerry Pournelle (1985) features an Orion-based spacecraft used in a last-ditch effort to defend Earth from invading aliens. As

Interstellar Rocket Technology

discussed by Arthur C. Clarke in his recollections of the making of *2001: A Space Odyssey* in *The Lost Worlds of 2001*, a nuclear pulse version of the "Discovery" was considered. A *Star Trek: The Original Series* episode, For the World Is Hollow and I Have Touched the Sky, features a generation ship, constructed out of a hollowed-out iron asteroid, propelled by "Orion class nuclear pulse engines" in which fission bombs are detonated in shafts. It appeared to have been traveling for 10,000 years and had travelled 30 light years on its own power. The 1998 film *Deep Impact* featured a spacecraft named Messiah which utilized the "Orion drive" and appears to be a variant of nuclear detonation propulsion. The novel *Anathem* by Neal Stephenson (2008) features an alien interstellar ship that includes a nuclear pulse drive. In the FOX TV series *Virtuality* (2009), *Phaeton*, Earth's first starship, is propelled by an Orion drive. The novel series *Troy Rising* by John Ringo, including *Live Free or Die* (2010), *Citadel* (2011) and *The Hot Gates* (2011), featured a series of extremely large spaceships using Orion drives as "the only practical method of maneuvering a multi-trillion ton platform". The premise of the 2014 mini-series *Ascension* was inspired by Project Orion and features a generation ship.

Other approaches such as solar sails will not work in interstellar space where sunlight pressure is nearly absent, so the entire acceleration phase (and deceleration phase) must occur close to a star. Only extremely large and massive sails the size of the entire state of Texas, would achieve the needed accelerations over the course of a month or less, and even these would not be much more than a few percent the speed of light at the maximum, making for very long journeys of centuries or more to get to the nearest stars.

Interstellar Travel

"*An extremely lightweight 62-mile-wide (100 kilometers) sail unfurled close to the sun could make an interstellar voyage in 1,000 years*", said Les Johnson, deputy manager of NASA's Advanced Concepts Office at the space agency's Marshall Space Flight Center in Huntsville, Ala.

There have also been frequent proposals for a laser-based system where a powerful laser near Earth shines on a solar sail and provides acceleration out to the limits of the solar system and beyond. The drawback is that although such systems may work for reaching the desired speed, there would have to be an equivalent system already installed at the destination to provide deceleration. Also, the pointing of the laser beam would have to be fantastically accurate to avoid losing power, and the laser power itself would exceed the entire energy usage of the world!

Interstellar Rocket Technology

In 1984, Robert Forward described an innovative laser-powered interstellar solar sail system that might actually work.

"This paper discusses the use of solar system-based lasers to push large lightsail spacecraft over interstellar distances. The laser power system uses a 1000-km-diam. lightweight Fresnel zone lens that is capable of focusing laser light over interstellar distances. A one-way interstellar flyby probe mission uses a 1000 kg (1-metric-ton), 3.6-km-diameter. lightsail accelerated at 0.36 m/s² by a 65 billion watt laser system to 11% of the speed of light, flying by Alpha Centauri after 40 years of travel. A rendezvous mission uses a 71-metric-ton, 30-km diameter payload sail surrounded by a 710-metric-ton, ring-shaped decelerator sail with a 100-km outer diameter. The two are launched together at an acceleration of 0.05 m/s² by a 7.2-trillion watt laser system until they reach a coast velocity of 0.21 c. As they approach Alpha Centauri, the inner payload sail detaches from the ring sail and turns its reflective surface to face the ring sail. A 26-trillion watt laser beam from the solar system, focused by the Fresnel lens, strikes the heavier ring sail, accelerating it past Alpha Centauri. The curved surface of the ring sail focuses the laser light back onto the payload sail, slowing it to a halt in the Alpha Centauri system after a mission time of 41 years. The third mission uses a three-stage sail for a roundtrip manned exploration of the Eridani at 10.8 light years distance."

The fundamental problem is that, based on what we now know about the location of interesting destinations, they occur much farther away that Alpha Centauri and Epsilon Eridani, so the power and time requirements for an Earth-based laser system become far higher, and increase as the square of the distance. A likely destination at 20 light years would require 25 times the power needed to get to Alpha Centauri or 180 trillion watts!

Interstellar Travel

Also, the mass being transported is only 1 ton for the sail and 2 tons for the spacecraft. For anything like a manned mission, the energy requirements, sail size and mass grow enormously when you add humans and their life support systems, and for likely destinations more than 20 light years away, become entirely unworkable.

Cosmic Pollution

One last thing we have overlooked in all of our great propulsion ideas is, what happens to the 'exhaust'? For our hypothetical spacecraft traveling at 20%C, we needed a kinetic power of 5.8 trillion watts during the speed-up and slow-down years. If we get this by accelerating protons to a modest 99% C each proton has an energy of 8.6×10^{-10} Joules/proton. This also equals a kinetic energy of 5.4 GeV per proton. As we saw earlier, the rate at which we eject protons from the engine has to be about (5.8 trillion J/s) / (8.6×10^{-10} Joules/proton) or 6.7×10^{21} protons/sec. That works out to about 0.01 grams per sec or 1100 Amperes! If our engine has a diameter of, say, 100 meters, this represents a flow of protons out of the engine at a rate of (6.7×10^{21} protons/sec) / (π (50m)^2 = 8.5×10^{17} protons/meter^2/sec. Needless to say, this would be a very lethal place for you to stand as the engines are operating full-bore, but what happens farther away?

This particle flow is not like a laser beam. Laser beams are composed of photons that do not interact with each other, so you can cram as many of them into a given volume of space as you like. That's why the beams stay collimated over such long

Interstellar Rocket Technology

distances. But protons are like-charged particles and repel each other, so as they exit the engine as free particles, they will cause the emitted matter beam to start spreading out laterally in space. As they exit the engine and cover the 100-meter area evenly, their average distance from each other will be about 100 meters/$(6.7 \times 10^{21}$ protons$)^{1/2}$ = 1.2×10^{-9} meters. The electrostatic 'Coulomb' force between them will be $F = kq^2/r^2$ where k is the Coulomb constant in SI units which is 9×10^9, q is the proton charge of 1.6×10^{-19} Coulombs and $r = 1.2 \times 10^{-9}$ meters, so $F = 1.4 \times 10^{-18}$ Newtons. The acceleration on the charge is found with $F = ma$ where m is the mass of a proton of 1.6×10^{-27} kg, so a = $1.4 \times 10^{-18}/1.6 \times 10^{-27}$ = 8.7×10^8 meters/sec². After 1 second, the beam is still moving forward along the axis of the ship at 99%C, but at least at the start, it is expanding transverse to the beam at a speed of V = acceleration x time = 8.7×10^8 meters/sec, which is 3% light speed. So very approximately, the beam's opening angle is just 2xarctangent (0.03/0.99) or about 4 degrees. By proportions, by the time the protons get 1 light year from the ship, the diameter of the beam has increased to 1 light year x 3/99 = 0.03 light years. Since 1 light year is 9.5×10^{12} kilometers, the circular area of the beam is π $(.03 \times 9.5 \times 10^{12})^2$ = 2.6×10^{23} meters².

How many protons are flowing through this area? At the spacecraft there were 8.5×10^{17} protons flowing through every square meter per second. This area has now increased from A = π $(50$ meters$)^2$ = 7800 meters² to 2.6×10^{23} meters², for a dilation factor of $2.6 \times 10^{23}/7800$ = 3.3×10^{19} times. That means the proton flow has been reduced at 1 light year to about $8.5 \times 10^{17}/3.3 \times 10^{19}$ = 0.03 protons/meter²/sec.

Interstellar Travel

This doesn't sound like much, but cosmic ray protons at this same energy of 5.4 GeV, our graph of the cosmic ray spectrum in the section in this book about Radiation, indicates a flow rate of about 700 cosmic ray particles per meter^2 per second, so at a distance of 1 light years it will be impossible to distinguish your engine exhaust from normal cosmic rays. A more detailed calculation will give better results, but the ball-park answer will probably not change. Your exhaust will not be a hazard, or likely even detectable by anyone at a distance of a few light years! This is a shame, because we might have been able to detect alien interstellar engine exhaust just by looking for weird cosmic ray signals!

'Warp factor 4 Mr. Sulu!'

As we saw in our first chapter, ever since science fiction authors started proposing interstellar travel by 'warping space' this has been a meme we have not been able to shed in over 80 years! Einstein's theory of General Relativity was the mathematical statement that space is warped in the presence of matter, and from this you can get everything from black holes to the Big Bang. Although the theory has been fantastically accurate in modeling the world of extreme gravity, it is after all a mathematical theory.

Physicists are fully aware that the laws of physics are only a small subset of the possible statement of these laws with mathematics, and that all mathematical theories can be stretched to cover conditions that are unphysical in our world. There is nothing about General Relativity that says it is not subject to this same limitation of interpretation. Only the Real World and its quantifiable data are the guides to use, not mathematical predictions. Every prediction based on the mathematics of General Relativity must be checked, and compared with other predictions to insure logical consistency. There will never be a prediction that stands alone by itself within a mathematical theory, which does not logically connect with many others.

The idea that space (actually spacetime) can be warped artificially and that this can be used to travel faster than light is one of those odd predictions from General Relativity (GR) that has gotten huge press over the years. In 1994, Miguel Alcubierre, a Mexican theoretical physicist well-versed in GR found an unusual solution to the gravity equations of GR. The

Interstellar Travel

solution was a particular distortion in spacetime where space behind a point was expanding, while space in front of the point was contracting. The net effect was that the point was driven forward faster than light in the surrounding space, but all speeds were less than the speed of light inside the 'warp bubble' solution. This figure shows what this would look like.

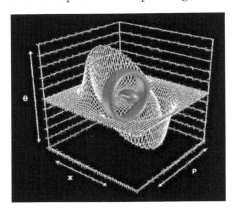

It was a purely mathematical solution. Physicists usually start from a physical object like a collection of mass, and then solve the equations to see what kind of distortion in spacetime results. That's how black hole solutions were first discovered. What Alcubierre did was just the reverse. He started from a desired distortion in space, then he worked backward to see what kind of mass he needed to make it happen. If that mass were, for instance, a spacecraft, you would instantly have a new form of space travel.

The problem was in the details. To make space dilate and collapse the way you needed in order to push a spaceship forward to the stars, you needed a lot of negative matter. Nothing like this exists except perhaps dark energy. The solution works in GR but as a mathematical solution it does not speak to real physics any more than you can directly see virtual particles and manipulate them.

Warp factor 4 Mr. Sulu!

This hasn't stopped some prospectors of this exciting 'mother lode' of ideas from pushing forward. Even NASA has a researcher, Dr. Harold White at the NASA Engineering Directorate in the Johnson Spaceflight Center, who is the Advanced Propulsion Theme Lead. He has found what he calls 'other loopholes' in the mathematical equations that say warping space might be possible. He is now trying to find experimental evidence that Nature actually allows these same loopholes. But we know that Nature has been very stingy in the past when it comes to giving physicists everything they want from a mathematical theory. Newton's physics work phenomenally well so long as you do not push them to high speed. Then the mathematics gives very precise, but entirely wrong, predictions for faster than light motion that is never observed, because light-speed is not a speed limit at all. But when engineers build things they have to obey the laws of Einstein not Newton. That's why our cell phone GPS system works.

Another idea that seemed very promising was the Bussard ramjet proposed in 1960 by physicist Robert Bussard. The idea was that space was filthy with gas; about 10^{-21} kg of hydrogen atoms in every cubic meter of space. If you collected this mass with a large 'electromagnetic' field thousands of kilometers in diameter, you could use it as fuel to run a fusion-based rocket that would propel you forward. The faster you went, the more mass you collected, so eventually you would accelerate to nearly the speed of light.

Science fiction authors caught wind of this idea and used it in many stories. popularized by Poul Anderson's novel *Tau Zero*,

Interstellar Travel

Larry Niven in his *Known Space* series of books, Vernor Vinge in his *Zones of Thought* series. Bussard ramscoops are also seen in *Star Trek*, where they are situated at the glowing tips of the warp nacelles of spacecraft and shuttlecraft, although the hydrogen is not used as nuclear fuel.

Since the time of Bussard's original proposal, it has been discovered that the region surrounding the sun has a much lower density of interstellar hydrogen than was believed at that time.

Robert Zubrin and Dana Andrews analyzed one hypothetical version of the Bussard ramjet design in 1985. They found that the ramjet would be unable to accelerate into the solar wind. The drag force would exceed the thrust of the ramjet for most plausible designs and ship speeds. The Bussard Interstellar Ramjet engine concept uses interstellar hydrogen scooped up from its environment as the spacecraft passes by to provide propellant mass. The hydrogen is then ionized and collected by an electromagentic field.

What about black holes?

OK. Let's go ahead and use black holes or worm holes like the magical Big Boy civilizations do in science fiction. Here are some unpleasant facts to pop that balloon.

Annoying Fact 1: Black holes are unavailable.

To use a black hole as a transport system you need to find one first. Astronomers are real good at that because black holes are

Warp factor 4 Mr. Sulu!

messy eaters. They swallow very hot gas and produce x-rays that can be seen across the light years. The nearest known black hole after several decades of searching our neck of the Milky Way is called Cygnus X-1 and is located about 6,100 light years away. To get to this black hole and use it for interstellar travel, you first need... interstellar travel!

Annoying Fact 2: You can't make artificial black holes.

Another option is to make your own black hole. The favorite method is to grab ahold of one of those microscopic quantum black holes at the so-called Planck Scale of spacetime and enlarge it to human or spacecraft dimensions. Then like a miner bracing the rock walls with timber, you shore it up with Kip Thorn's anti-gravitating matter to stabilize it. That's what the clever aliens did in *Contact* and in *Interstellar*, and perhaps even *2001: A space odyssey*. But there is one big problem. Our spacetime doesn't look like the fashionable 'quantum foam' model that everyone likes to use as a starting point. There have been several different searches for evidence of a grainy quantum shape to spacetime at the Planck Scale of 10^{-33} centimeters. It doesn't seem to be there!

Since 2009, astronomers such as Robert Nemiroff have used the photons from gamma ray bursts to see if the higher energy photons traveled a longer distance through space than the lower-energy ones. If spacetime is foamy, then the higher energy photons will be scattered around a bit more like the balls in a Pachinko game compared to the slightly more direct route taken by the lower energy photons. Well, several attempts to measure

Interstellar Travel

the arrival time differences have turned up Nada! Zippo! Nothing! Also, folks have looked at the polarization changes which should also take place and also get null results at a level hundreds of times smaller than the Planck Scale.

So, there is no evidence that John Wheeler's visually evocative, quantum spacetime foam model from the 1960s with quantum black holes exists or is in any way an accurate way to describe what the bedrock-scale of spacetime looks like. It could be stringy, or loopy but not frothy. That means Mother Nature won't let us inexpensively borrow some already-existing phenomenon in our universe to co-opt into a transportation network. Even worse, those folks that work in string theory and M-theory say that our 3-brane universe is located in the 11-dimensional Bulk where only gravity can travel from one 3-D space to another. The very particles that make up our bodies and starships cannot travel outside of their 3-D brane spaces. These assertions from Brane Theory have yet to be tested!

Annoying Fact 3: Black holes will kill you.

The only way that Nature can make black holes today is through stellar evolution and the supernova mechanism. Supermassive black holes are created through accretion of matter as they grow, but the nearest of these is 27,000 light years away so you need interstellar technology to get there before you can use it for transportation... an interesting tautology!

Anyway, stellar-sized black holes are lethal because with a horizon only a few tens of miles in radius, gravitational forces will pull you and your ship apart before you got within a few

Warp factor 4 Mr. Sulu!

thousand miles of the horizon. You would never get close enough to the event horizon to go inside and use it to short-circuit space. If someone reading this has a good way of how to negate the Inverse-Square Law of Gravity, or the Inverse-Cube Law of Tidal Forces which will both kill you, then go for it.

Annoying Fact 4: Super civilizations are no help

Many science fiction stories like to jump-start their tales of galactic exploration by using the *deus ex machina* of an advanced civilization giving us interstellar travel as a gift , or leaving it behind as they went on to other adventures. There is only one set of fundamental physical laws per universe. If not, our universe would exist in chaos where one region of space had different laws than another. Astronomers would EASILY see this effect, and so could you by just looking at the night sky! So, advanced civilizations have to play by the same rules that we do. You might fantasize that they know something that 21-st Century scientists don't, but I would not bank on it. Besides, how long do you want to wait before some aliens hand over what is probably one of their most important secrets? What would be the cost they would exact from us in exchange?

When you look at interstellar travel this way, it's like Dorothy said in the Wizard of OZ...*There's no place like home*... and for us it is a solar system with a huge amount of real estate and enough cool things to do to occupy us for millennia! By the way, instead of getting to Alpha Centauri in a century, could you possibly get me to Mars in a few days? The *Astronomy Café* over at Elysium has a mocha latte that is to die for!

Interstellar Travel

Magic and Advanced Tech: The Big Lie

The way you look at the future and the prospects for interstellar travel depends on whether you think there are actual limits to what humans can accomplish, or whether no limits really exist given enough time and motivation. If I can imagine taking a swim in the liquid-iron oceans at the core of our Earth, does that mean that 'someday' we will have the technology top do that? Would I be a Quitter and a narrow-minded, uninspired scientist if I said '*No we will categorically never be able to do that. ?*

The Big Lie comes from Arthur C. Clarke's Three Laws, which are:

1. When a distinguished but elderly scientist states that something is possible, he is almost certainly right. When he states that something is impossible, he is very probably wrong.

2. The only way of discovering the limits of the possible is to venture a little way past them into the impossible.

3. Any sufficiently advanced technology is indistinguishable from magic.

Interstellar Travel

A very large community seems to feel that these purported Laws written by a science fiction author are in some sense the missing Commandments brought down from Mt. Sinai by Moses. The Third Law is especially interesting, and it's not even unique to Mr. Clarke. It first appears in his 1973 revision of "*Hazards of Prophecy: The Failure of Imagination*". But it resembles very closely a statement in a 1942 story by Leigh Brackett: "*Witchcraft to the ignorant, ... simple science to the learned*". Even earlier examples of this sentiment can be found in the story Wild Talents (1932) by Charles Fort, "*...a performance that may someday be considered understandable, but that, in these primitive times, so transcends what is said to be the known that it is what I mean by magic.*"

The implicit lie behind the Third Law is the well-known fallacy of circular reasoning. The word 'sufficiently' is the red herring and begs the question 'Sufficiently relative to what?". Because the Law provides no definition for what is meant other than asking the reader to fill-in what they believe it means, we have to test sufficiency against the idea that the technology be advanced enough to resemble magic. We then have the tautology 'Any technology advanced enough to look like magic, is indistinguishable from magic'.

This is not very helpful!

It is usually interpreted to mean that if we brought a radio to an ancient Egyptian Pharaoh it would appear magical to him, or if we brought a cell phone to an isolated Amazon tribe like that Yanomami, its workings would also appear magical. Is this really the correct comparison? For many ideas in interstellar travel, some technological revolutions will be required in the next few

Magic and Advanced Tech: The big lie!

centuries. But if we go back to the 1800's, we could surely explain the workings of a cell phone to James Clerk Maxwell who invented the mathematics of electrodynamics, or describe our sophisticated chemistry and basic quantum physics to the users of the first spectroscopes in the 1850s. The fallacy of the Third Law is that it makes you believe we are stupid 'savages' who are clueless about the workings of the physical world, and we have absolutely no way to go from what we know now to a future breakthrough without invoking Harry Potter's magic. As you probably understand by now, this is a ridiculous comparison. We really do understand the physical world pretty well.

There is nothing about the 'magical' rocketry of the 21st century that a physicist could not have anticipated from what was known about the physical world in the 1800s. This is when Newton's Laws were in good shape, we understood how charged particles move in electric and magnetic fields, and the majority of the Periodic Table was known. Chemical rockets, even ion engines, could have been directly surmised in back-of-the-envelope calculations by late-19th century physicists. Even travel at 20% the speed of light could be pretty accurately treated by Newtonian physics, as well as the likely travel times to nearby stars known at that time.

Today, there is even less wiggle room for throwing out the known laws of physics just to get to the stars. Physicists have now looked into so many different corners of the world, even our artificial attempts at violating these laws leads nowhere. It would be exciting if they did because that would start a whole new round of experimentation and thinking 'out of the box' to

Interstellar Travel

understand what just happened! Special and general relativity, along with quantum mechanics and chemistry have been tested sometimes to ten or more decimal places looking for some deviation, and only our apparent current mastery of these laws is reflected back at us. If Star Trek's new physics is 'out there' it is not a kind of physics that extends from what we know and have validated today. What is worse is that what we know today is such a terribly interconnected lattice of ideas, evidence and validations. To sneak something new into the mix would be a lot like adding a new chip to a 1000-piece puzzle and expecting it to make flawless sense in the whole.

So, I would not bank on Clarke's Third Law to bail us out. If we want to get to the stars, it will be from a clever technological extension of the laws of physics we already know, and the basic idea is to throw stuff out the back of your ship as fast as possible to move it forward. That takes both energy…and money. Lots of it!

Where is everyone?

In 1950, Nobel prize winner and physicist Enrico Fermi asked a question that still resounds in the halls of, not only the search for extraterrestrials, but also among the many designers of interstellar travel technology" *Where are they?*

He was having lunch, as the story is told, with Edward Teller and Herbert York and they were discussing life in the universe. The problem he saw in an instant of insight was the paradox that, although there might be many intelligent civilizations in the Milky Way galaxy, and it only takes one civilization a few tens of millions of years to visit all the stars in the Milky Way, we have no credible and obvious signs of any such visitation, nor do we seem to have any astronomical evidence for such civilizations.

In the 65 years since Fermi's question went public, there have been many discussions of this paradox, and with programs such as Ozma and SETI, quite a few sophisticated radio searches for any signs of artificial signals from space. Despite a few exciting false alarms, there is no evidence for a Milky Way populated by advanced civilizations, at the same time that astronomers are increasingly certain that life, at least bacterial forms, is probably ubiquitous.

Many of the counter arguments opt in favor of our solar system being embargoed by some vast and unprovable conspiracy of super-civilizations who have erected No-Go Zones around this portion of the Milky Way. Others favor various flavors of a statistical argument such as intelligent civilizations are rare

Interstellar Travel

because they self-destruct, or that every last one of them has grown disinterested in communicating with primitive life (us).

The cheapest way to make contact is through some form of radio or optical communication. It is far, far less expensive to send photons into space than rockets. This is exactly why many generations of radio and optical searches have been attempted over the decades. You can still participate in one of these searches yourself by joining the *SETI@home* program. As of 2010, after 10 years of data collection, *SETI@home* has listened to one frequency, 2.5 MHz, using the Arecibo radio telescope in Puerto Rico, and has surveyed about 20 percent of the entire sky. Over 180,000 people using 290,000 computers around the world volunteer to analyze chunks of the signal, and use established protocols to check for unusual signals. None found so far.

Between 1995 and 2004, Project Phoenix concentrated efforts on that component of the NASA SETI project known as the Targeted Search. Its strategy was to carefully examine the regions around 1,000 nearby Sun-like stars. The world's largest antennas were used, committing observing time for SETI. Project Phoenix was orders of magnitude more comprehensive than any experiment yet performed. Over the span of a decade, the project examined about 800 stars within 200 light years. The search was sensitive to the equivalent of strong terrestrial radars, but found no evidence of extraterrestrial signals of that strength.

This radio silence could just mean that intelligent life elsewhere is broadcasting on some other frequency, or that their equivalent of our television and radio programs use more efficient methods

Where is everyone?

like fiber optic cables. An advanced civilization only spends a very short time broadcasting radio energy wastefully into space before it starts to use more efficient means. We are already at that point with fiber optics and highly-targeted satellite transmissions. But surely if there are as many civilizations in the Milky Way as we hope there are, at least a few of them would still be primitive enough to still be leaking out radio energy that we could hope to eavesdrop upon. The silence we hear from surveys out to 200 light years suggest we live in a 'quiet' part of the Milky Way. We can turn this around and already place an upper limit to the number of radio-emitting civilizations in the galaxy that could still be out there.

The volume of the Milky Way can be approximated by a disk with a thickness of 1000 light years and a radius of 50,000 light years.

$V = \pi R^2 h = \pi (50,000)^2 (1000) = 7.9 \times 10^{12}$ cubic light years.

The volume of a 200-light year-diameter sphere is
$V = 4/3\pi (100ly)^3 = 4.2 \times 10^6$ cubic light years.

So dividing one by the other we get no more than 2 million civilizations otherwise SETI would have seen something in its 200-light years search! The hope is that with more sensitive and comprehensive radio searches to distances of thousands of light years, we will either finally hit upon the first artificial signals, or we will continue to whittle the number of civilizations transmitting today from 2 million down to a few hundred…or less. Then the Milky Way starts to look more like an intellectual desert than a vibrant and thriving community of advanced species.

Interstellar Travel

Another interesting argument that comes up from time to time is that, putting aside the communication silence, interstellar travel may itself be very difficult. Perhaps so difficult even robotic probes cannot be designed to weather the journey, let alone spacecraft carrying the most earnest and motivated aliens.

Any probe that inadvertently leaves its own solar system quickly becomes as dysfunctional as our own Pioneer and Voyager spacecraft are today, and impossible to detect at light-year distances. If it were possible to make the trip, many technologists believe hardy robotic explorers will consist of some form of self-replicating von Neumann factory. The complex, low-mass probe would enter a remote planetary system, find an asteroid and start to mine it. Soon it would be using its stored blueprints to fabricate a large installation capable of communicating with Homeworld across the light years, sending newly-minted robotic probes to collect data from around its new planetary system, and then creating new von Neumann probes to continue the journey to other stars. What would such a probe look like in our solar system? Would we even know it existed?

In one intriguing 2001 science fiction story by Stephen Baxter called *Manifold Space*, the evidence for many waves of these probes invading our solar system every few million years is actually hidden in plain sight! The entire crust of Mercury had been removed to provide raw material for additional probes, leaving its denser mantle and core behind for astronomers to puzzle over. Other visitors de-spun the Earth-like world of Venus and mined it ruthlessly for raw material to fashion new

Where is everyone?

ships. Are the rings of Saturn actually the result of alien, robotic mining eons ago?

The numbers are compelling when you consider the possibilities. The results from the Kepler survey alone suggest that one in five stars have planets like Earth that have temperatures suitable for liquid water. In a galaxy with over 1 trillion stars, that works out to 200 billion planets with the same mass and temperatures as Earth. Now the only question is, how many of these have biospheres that evolved advanced life forms that we could communicate with?

Unfortunately that is the hardest part of the equation to solve. The first thing we need to discover is, how common are atmospheres that have free oxygen: a dead giveaway for an active biosphere. We are only just able to study the chemistry of a few planets so far and there is much more to do to determine how many of the 200 billion planets have biospheres. After that it is a question of what fraction of those evolved intelligent, technologically-savvy creatures?

With the SETI program, it is hoped that we can eventually detect leakage radiation from such alien civilizations. Other than that, there are no other ways to decide from looking at a planet from thousands of light years away whether it has intelligent life or not. The only possibility is to look for city light emission on the 'dark side' of the planet, but how does one distinguish that from other sources of planetary light such as aurora?

Interstellar Travel

The Pessimist's View

I have been an avid science fiction reader all my life, but as an astronomer for over half my life, the essential paradox of my fantasy world can no longer be maintained. Basically, science tells us that traveling fast enough to make interstellar travel possible requires more money than society will ever be able to invest in the attempt.

Einstein's theory of special relativity works phenomenally well, with no obvious errors in the domain relevant to space travel. His more comprehensive theory of general relativity also works exceptionally well and offers no workable opportunity to "warp" space in a way that can be technologically applied to space travel without killing the traveler, creating a mathematical fiction of 'negative matter', or having to generate enough energy to incinerate the universe. Interstellar travel will be constrained by the reality of special relativity and general relativity, and there is no monkeying with Mother Nature to make science fiction a reality.

So the only way to get from here to the nearest stars is by the slow-boat method of some kind of kinetic or radiative propulsion. There are many workable ideas, such as ion drives, fusion drives and solar sails. In fact, the all-around best ready-to-go idea is the ion engine, which is an off-the-shelf technology that has been used on many satellites and several spacecraft so far. With a small but constant thrust applied over months, years and decades, scaled-up versions of these systems could boost small payloads to over 10 percent of the speed of light in a few

Interstellar Travel

years, allowing travel times to nearby stars of as little as a century or less.

All plausible interstellar missions require travel times of centuries not decades, which means the complex technology must be 100-percent reliable and/or self-repairing. There can be essentially no moving parts, because friction would create wear and tear over decades and centuries of use.

It is entirely plausible to think about kilogram-sized payloads that can be boosted to near-relativistic speeds very economically, but this is impractical because at the distance of the nearest stars, you need a powerful and massive transmitter that can relay data back to Earth, or what is the point of the journey? Even a Voyager-class spacecraft with, say, a "souped-up" megawatt radio transmitter could not be detected at Alpha Centauri by the largest Earth-based telescopes, even at data rates of 1 bit per year! A laser-based system would be highly directional and could possibly do the trick, but it would weigh tons, not kilograms.

One option could be a sophisticated nanotechnology von Neumann-based system with a mass of a few dozen kilograms that would arrive at its destination, find an asteroid to mine, and then build from scratch a much more massive system capable of carrying out the scientific investigation and relaying the data back to Earth.

But the concept of sending humans to the stars makes no sense technologically, or at an economic scale that would interest humanity as it is currently constituted. Even if we were at the brink of extinction, do you really think that 7 or 10 billion

The Pessimist's View

humans would want to foot the bill and the decades-long effort to send a few lucky humans on a one-way trip to a distant planet -- that may not even be habitable?

Most people severely underestimate how big the interstellar frontier is. Not even astronomers who study this 24/7/365 as part of their careers have a good gut instinct for the distances involved. So any technology we have today, or any extension of it that we can actually build, will not match the challenge of interstellar travel, especially at a price human society will care to pay. Humans pay for big projects that have intermediate goals that are also worthwhile. The railroad system or highway system could be used within a few months for limited but expanding service. There are no such intermediate goals for interstellar travel that have social benefit and can keep the momentum for funding going. It's all or nothing. Every technology put forward has major drawbacks in implementation and human safety. Even unmanned probes are expensive and will vanish from human memory by the time they reach their destinations centuries after launch!

What is the point of investing trillions of dollars to arrive at a solar system where you still have to build sealed buildings and wear spacesuits as a colonist? You might as well live in an orbiting habitat! The true destination for humanity's first starship will not just be a habitable-zone-Earth-sized world but a world with a detected biosphere producing a breathable oxygen atmosphere. We do not know the average distance to such a world. A sphere centered on our sun with a radius of 300 light-years would encompass about 500,000 stars. That would be enough to have about 100,000 Earth-sized planets in habitable

Interstellar Travel

zones. Do you think that 100,000 planets is large enough to sample to find a single worthy target with a biosphere-generated, breathable atmosphere?

The point is that, before we make an expensive interstellar journey, we will already have cataloged the possible targets and know about their properties, sizes, liquid water and atmosphere composition long in advance. It is much cheaper to gather this information before the trip than mount an expensive "shot in the dark" expedition hoping for a good outcome. We do not operate that way today, so why would we do so in the future with a far more expensive mission?

In another few decades, we will know the answer to the question of where the likely target stars are for an interstellar journey. This will also, by the way, answer the question "Is there life elsewhere in the universe?" They will not be close-by like fabled Alpha Centauri or Barnard's Star. They will be remote (hundreds of light-years) and put even greater burdens on transit times and signal transmission back to Earth.

The Optimist's View

Forget star travel! Let's get busy and really have some fun with our own back yard first!

Our solar system is vast, and a big-enough playground for human exploration to last us for centuries. It is technologically accessible to us even today, as the numerous unmanned spacecraft and robotic systems clearly show. There are many scenarios that can be planned over decadal or century timescales that would have human outposts and colonies on just about every interesting body in the solar system, from planetary surfaces and the surfaces of their moons to asteroids and comets.

But is manned exploration the only way to go for now? Absolutely not! When you subtract manned exploration, which is hugely expensive, and replace it with robotic rovers that relay high-definition images back to Earth, all of humanity can participate in their own personal and virtual exploration of space, not just a few astronauts or colonists. The Apollo program gave us 12 astronauts walking on the lunar surface, a huge milestone for humanity, but today we can do the Apollo program all over again and augment it with a virtual, shared experience involving billions of people! This is the wave of the future for space exploration, because it is technologically doable today and scalable at ridiculously low cost per human involved. NASA's Curiosity rover is only the Model-T vanguard of this new approach to human exploration. More sophisticated versions will eventually explore the subsurface ocean of Europa

Interstellar Travel

and the river systems on the "Earth-like" world of Titan -- perhaps by the end of this century!

I know this robotic vision of human exploration doesn't match up with the Star Trek or Babylon 5 versions of the future, where flesh-and-blood humans explore the galaxy in starships. As an avid science fiction reader, I too am disappointed that we live in a universe where interstellar travel seems permanently beyond reach in any kind of human future that makes scientific or economic sense. But this is the deck of cards that we seem to have been dealt. We can pine away for a mythical future of interstellar colonization, but that will be a reality for a future humanity that looks nothing like our civilization, perhaps driven by extinction to help focus the resources toward that goal.

Meanwhile, if you want any kind of space exploration that matters within the next century or beyond, it will be robotic, virtual, and involve billions of people, not just a few very lucky travelers -- so what's wrong with that?

We could also invest in nanotechnology to launch interstellar probes weighing only a few kilograms but capable of building, at the destination, thousand-ton laboratories for remote exploration and returning virtual-reality information back to the Solar System for all of us to enjoy, not just a few colonists. There has yet to be much science-fiction literature that thoroughly works out the possibilities of such a solar system-girding civilization.

The Optimist's View

So there you have it. In the next few centuries, we can colonize the solar system in any number of different ways using largely conventional technology extended to meet the reasonable challenges of week-long hop to Mars, Saturn or elsewhere. Without violating any known physics, ion engines already used on NASA's Dawn mission and many commercial satellites, along with solar sails combined with nuclear power plants (developed by NASA since the 1960s), can probably be scaled up to meet these challenges.

We can terraform the whole solar system if we choose, and create from it the economic and technological infrastructure that may someday make interstellar travel a desirable goal. We can even create nanotech unmanned missions and flood our local neighborhood with an expanding network of information-gathering probes. We can build remote imaging systems capable

Interstellar Travel

of mapping the surface of a world 500 light years away with near-photographic detail. This can be done at a fraction of the cost of manned travel and, by virtual reality, involve billions of people in the adventure. So, let's do this next step right and, for now, not waste our energies trying to run to the stars before we can walk in our own backyard!

Meanwhile, we should keep reading books like Gordon Dickson's *Mission to the Universe* and dwell on the book jacket synopsis: *"Benjamin Allen Shore is in Command of Earth's first starship. But it is an empty command: The President of the United States has grounded the pulse ship indefinitely. Shore knows there are planets suitable for human habitation, and he knows his ship can find them. With carefully forged orders and an untried crew he embarks on an awesome journey — to the heart of the galaxy. What he finds there is more than he ever hoped for...the stars have worlds and people. And there is no way he can be prepared for the menace of the Golden People — an ancient people who have driven all other races from the Galactic Center — or for what awaits him if he ever returns to Earth."* Remember, as yet, there is no scientific way to prove this story wrong, at least in its basic themes.

Conversations with the Public

In 2014, I wrote a number of blogs for the Huffington Post that became very popular and produced hundreds of comments. Here are some of the best of them and my responses in italics.

Saying we as a species will never travel out of the solar system is absurd. A million years from now do you think we would have the same technology we do today? *I agree. It would be absurd to make any predictions about humanity even 500 years in the future. But most folks who discuss interstellar travel claim it will happen within a few centuries. On that timescale we have a lot more evidence about human behavior, goals and technology. You and I and the circumstances under which we live are not that different than folks living 150 years ago.*

Why create artificial environments on nearby planets? Why live indoors? I can't imagine a more miserable existence than living in one of those Mars One pods. *This is one of my points exactly. Interstellar travel will not be sold to humanity if it means visiting a planetary system where we cannot travel to the surfaces of the planets because of their gravities, or live inside a sealed habitat because of their noxious atmospheres. Our destination will have to be so nearly Earth-like that the chances of finding one of these will be improbable within the reach of the transport technology.*

Within the next 100 years, we may see colonies in space (like Gerard O'Neill forecast) and then the economics of interstellar travel will change

Interstellar Travel

dramatically in its' favor. Unfortunately, making interstellar travel a reality is a lot more than just finding the money. Even massive habitats in our solar system, or orbiting Earth, will force us to deal with such hazards as the collapse of the human immune system, psychological issues, collisions with meteorites or orbital debris, and what to do when complex technology fails.

Then again, maybe what we think we know about the laws of physics is actually wrong, and inexpensive FTL travel is possible. There is no government on Earth that would invest development money at the scale needed to develop interstellar travel, based on the unsubstantiated belief that our current understanding of physics has an imagined loop-hole favoring interstellar travel. A few hundred thousand dollars may be found to form study groups from time to time, but government always takes the lead for established science when deciding whether to fund a program.

"Everything is theoretically impossible, until it is done." - Robert A. Heinlein. My complaint is that science fiction authors say things to sell books, and so they have a huge vested interest in coming up with phrases, like Clarke's Three Laws, that sound good, but lack much meaning. In fact, the comments are usually illogical. This quote is illogical. If you want to base your view of the future on a statement, why use one that is illogical?

I think we should fix our problems here on Earth before propagating our racism, homophobia and hatreds throughout the cosmos. So, given that these problems have been part of the human condition for thousands of years, what you are saying is that we should never leave our planet until everything is perfect. That time will never happen, in fact exploration has always been

Conversations with the Public

the activity that brought humans into contact with new ideas and solutions to long-standing problems.

Two hundred years ago we lived without light bulbs, or anything else dependent on electricity. Those alive at that time would have found imagining a world with electricity very difficult, as we alive today find imagining the mechanisms that may one day enable interstellar space travel very difficult.

These are not the same things at all. The laws of physics we know today are the same ones that worked two hundred years ago, or two thousand years ago. Even though our ancestors did not know about them they still lived lives that were controlled by them. Laws of physics are not just a bunch of disconnected statements about the world. They are a highly integrated network of constraints on how matter operates. If there were a workable and inexpensive way to travel quickly to the stars, we would have seen signs of this in the way that high-speed matter works. We see no evidence anywhere in the universe for any large-scale violation of relativity or quantum mechanics that would enable interstellar travel.

Similar arguments were made by respected scientists and engineers against flying machines and space flight. "Heavier-than-air flying machines are impossible." — Lord Kelvin ca 1895. "To place a man in a multi-stage rocket and travel to the moon, I am bold enough to say that such a man-made voyage will never occur regardless of all future advances." — Lee DeForest, American radio pioneer and inventor of the vacuum tube, in 1926.

This kind of argument comes up over and over again in discussions about interstellar travel. But again, none of these ideas were contrary to the laws of

Interstellar Travel

physics known at that time. With interstellar travel, we are dismissing what we know about the laws of physics we know today, let along the 'laws of human biology and psychology, and trying to claim an equal certainty that we can make interstellar travel happen by using some other set of laws we have not the slightest evidence actually exist.

The entire reason for interstellar exploration in the first place is precisely BECAUSE we need a closer look at some of these systems because we don't know the nature of their planets.

This is not how modern space travel works at all. Because of the expense, governments demand that the financial investment be secure and that the effort have a huge probability of success. The Apollo program was possible because we knew before-hand that the moon existed and had a surface you could actually land upon. Interstellar travel will only be attempted if we know before hand through far less expensive means that there is a habitable planet with a promising non-lethal atmosphere at the destination.

You make this catalog of planets around other stars sound very convincing, but I don't believe we actually know for certain what's circling the nearest stars. We will not begin an interstellar voyage unless we already know a suitable planet exists to visit. It is not a matter of belief, but hard science and detailed observations that will tell us where these planets are located.

Its people like you who stunt humanities progress to push ourselves to understand the universe we live in. Scientists have given humanity all of the 'miracles' that now constitute the world we live in. The progress of science is all about individuals thinking outside the box.

Conversations with the Public

My comments about interstellar travel are entirely consistent with the rich scientific tradition of questioning our assumptions. Sometimes those questions are not very fun to consider, and the current discussion about interstellar travel suffers from too much fantasy thinking and not enough realism.

Given the scale of energy and material resources needed, to say nothing of the time, is it possible that no civilization has ever ventured very far from its birth planet, which would explain why we haven't been visited by aliens. This is a very good point, and the simplest on that explains the so called Fermi Paradox. If there are so many intelligent civilizations in our galaxy, why do we have not a smidgeon of evidence they exist?

Given the very real problems we have here on Earth, won't the exploration of Mars and the outer planets over the coming several centuries will prove daunting enough?

Yes it will, but there is nothing about this expansion into the solar system that violates the laws of physics. It is strictly a matter of national priorities. Unfortunately, as world budgets come under pressure to help aging populations and deal with climate change, there will be even less stomach for human exploration beyond lunar orbit.

The problem with much science fiction, and with the thinking of those who optimistically think the future will unfold in some manner like a science fiction story, is that it ignores the true scale of resources, energy, and time involved in bridging the distances involved. This is my point exactly. People continue to grossly underestimate the costs and impacts upon human health that come from prolonged journeys in space. These are rooted in the genetic makeup of our species, and to try to defeat them with clever technology often

Interstellar Travel

leads to other consequences even more costly to deal with. To cure the decline of our immune systems in space you need gravity and a dirty environment, both of which are expensive and have their own technology impacts.
If we remain confined to this single world, we will stagnate and devolve. But there are plenty of places to go in our own Solar System. I completely agree. Humans confined to the same environment year after year literally go crazy from boredom or develop other psychological adaptations. Just look at the kinds of behavior issues that have emerged from our over-populated world and our repetitive behaviors.

It's hard to envision interstellar travel when it takes hours just to commute to and from work on the Interstate.

I know what you mean! The speed change from 5 mph (0.0022 km/s) to interplanetary travel (10 km/sec) is a factor of 4500. To reach reasonable interstellar speeds we will need another factor of 6000 above current interplanetary speeds

The author of this book should watch more science fiction if he wants to speculate about the future. I do.

I have been an avid SF reader since my teenage years and the stories I have read have always been inspiring and stimulating. But just as readers of Harry Potter see this as entertainment and not a blueprint for our actual future, as a scientist I see the primitive science used in these SF stories as irrelevant statements about the real world. No SF story I have ever read even gets the physics we already know correct. Case in point, you do not need to fire your engines continuously to maintain your speed. Also, electrons do not have solid surfaces.

Conversations with the Public

For all we know the space between stars is a vast debris field. Hitting a space rock the size of a baseball that could be going 500 km/second could bring a quadrillion dollar project to an abrupt end. There doesn't have to be much of it.

This is quite right. It scares me to think that we can detect interstellar dust grains that are merely an annoyance to our ship and travelers, but there is no way to detect golf-ball-sized particles we could encounter every few years with devastating consequences. Even on the ship, our avoidance systems would not see these coming until the radar reflections saw them close enough that there would not be enough time. Then again, these may not even exist, but we have no iron-clad guarantees for the safety of our mission that we can fabricate from Earth.

Sorry but I remain a skeptic about the possibility of human flight to other planets. It's not the distance, it's not the cost or even the habitability of another world. We humans have been killing each other on this little blue marble we currently live on and we'll continue to do so until the human species no longer exists. How are we going to move on to other worlds when we can't seem to find some equilibrium and tolerance for each other right here in our own backyard.

This definitely is a problem. What kind of legal system and police force do we have to set up even during interplanetary voyages to deal with crewmates that have become a hazard to themselves and others? Even with as few as two people you could have one of them 'snap' and become a hazard.

Regardless, there's no question as to whether or not we will eventually need to go interstellar. We'll need to leave one day. Hopefully we have an answer by then.

Interstellar Travel

I would hope so too. The longer we have humans on only one location in the solar system, the more vulnerable we are as a species to dinosaur extinction events. On the other hand, it is cheaper to create a planetary defense system to ward off extinction-level asteroid impacts than it is to build colonies outside Earth's orbit.

Current mainly Einsteinian theory dictates that the speed of light is a limiting barrier as indeed with current technology it is. Just as the jet engine and the quantum jump in aerodynamics short circuited the sound barrier.

These are not the same things at all. The so called 'sound barrier' is not a physical barrier to faster speeds. It is just a statement about the aerodynamics of flight in our atmosphere. Relativity on the other hand works in our atmosphere and in the vastness of empty space and does actually set a speed limit for matter and energy that is consistent across many scales both atomic and cosmic.

Another reason why such a trip will not be possible is the concept of entropy. With time stuff will fail and will likely not be repairable. Thus, a trip to a star system 100 light years away might take a thousand years, will the systems on board work continually for that long?

This is a very important point and one that very few people think about because it is not as sexy as coming up with fantastic solutions to the kinds of propulsion technology needed.

Obviously, we will do the same thing we did with the Moon when it comes to interstellar travel....send probes first. Worrying

Conversations with the Public

about HZ planets and atmospheres is getting WAY ahead of ourselves.

Not true. We sent probes to the moon, Mars and other solar system bodies because we could do so for only a few hundred million dollars. To send any kind of probe to the nearest stars requires communication technology that even by itself weights many tons. There is no way to miniaturize a 100-meter radio communications dish or a laser system let alone probes to gather data that will not currently cost trillions of dollars and require centuries to get the data back to Earth.

What are the ethics and protocol for exploring such a world? Releasing a single bacteria on that biosphere could have catastrophic consequences for the existing life there.

This is a very good point. Any place we find with a breathable atmosphere will have a biosphere, but we have different populations of bacteria and symbiotic organisms that could be devastating to distant worlds. Today we even worry about polluting Mars with the very organisms we hope evolved independently on Mars, but if we find tem the specter that they were brought there by our Mars rovers is a real scientific concern.

We need to explore beyond our little world, what would Columbus say if he heard so many afraid to explore?

We are not afraid to explore, we simply want to make certain that there really is something at the other end of the very expensive journey worth seeing to recover the cost and risk. Columbus and Leif Erikson could use off-the-shelf technology to get to the New World. With interstellar travel, we do not have that option.

Interstellar Travel

We most definitely have the science to develop faster than light vessels using the work of Miguel Alcubierre on warp engines and theory.

This is an interesting use of Einstein's relativistic theory of gravity and spacetime, but as an astronomer, we have not seen a single object in the universe that 'warps' spacetime the way that Alcubierre proposes to do. All other human technologies are based on principles we have seen nature demonstrate for us to mimic in a different way.

It is NOT the science that is the problem. It's the economics. That we theoretically could do something is not the same as saying that we will do it, and it may very well be the case that the effort would involve the applications of resources on a scale unknown to humans to achieve.

That is the main point behind this book. Ultimately, humans will prioritize doing interstellar travel based on its standing relative to other economic interests. If every human on Earth wants to live like wealthy Americans in the 'upper middle class' we will need at least one or two additional Earths supplying the raw resources to support these needs. Interplanetary travel and asteroid mining might make economic sense, but interstellar travel will not.

We have a lot of advances to make before this even becomes feasible. IF it is even possible. One thing is for sure. If there is a way to do it the human civilization will figure it out. And if there is no way to do it, what will humanity become? I think those of us who are scientists, engineers and SF fans would find such a limited world very confining after a while and our will to innovate will have to find some other goals in life.

Conversations with the Public

What happens when we get there and find out the place is already teaming with life and they (like us) wouldn't be too excited to have the Jones move in next door to them.

This is a very good point, but the odds are against being so lucky on our first voyages. Everything is a matter of numbers and probability. If only one out of ten planets that are otherwise earth-like have oxygen atmospheres, we can see from the history of Earth that even when evolution took the 'right' path to intelligence, it took over 500 million years to get to homo sapiens 50,000 years ago. The probability that on our first try we encounter such a planet is very remote, and for Earth it is a 50000 years/ 5 billion or one chance in 100,000.

The ONLY way we are going to the stars is through our imaginations, the way science fiction writers have been doing it for years. Stick with Star Trek, it makes it easy.

This is probably true, but perhaps human motivations will change in the future in reaction to some existential threat. Then even hollowing out an asteroid to build a 100-generation colony ship may seem like good insurance. The problem is that contemporary humans never think about the survival of our species as any kind of important issue. In the future that may change.

Interstellar Travel

Epilog

The vast majority of our discussion has not even mentioned a technology for actually building an interstellar spacecraft, and already the number of problems seems out of control.

What should be our destination? What will we do when we get there? How large a crew will we need? What will happen to human health along the way, both physical and mental? What will we do when things break down? How much energy will we need to reach a reasonable interstellar speed, and what will we do about the interstellar matter we encounter?

Without even discussing which particular engine technology we will use, the answers to these very basic questions do not seem to have easy or inexpensive answers. Even some of the most promising solutions such as hibernation and microgravity are fraught with significant and seemingly insurmountable challenges.

But perhaps the biggest question we should all be answering is, what kind of life will the travelers have after reaching their destination? It is easy to say that they will journey down to the planet's surface and explore, set up a colony and begin the hard but exciting task of building our first interstellar community, but the reality is very different. It is likely that the planet will not be an exact twin to Earth in terms of atmospheric pressure and the beneficial 20% of free oxygen. It is statistically far more likely that it will have a lethal trace gas like cyanide, or offer far less than 20% free oxygen. That means our colonists will have to

Interstellar Travel

permanently live under a dome, and explore their planet in spacesuits for the rest of their lives there. Is this an acceptable quality of life after a potentially hundred-year journey?

Our likely destinations are also surprisingly few out to at least 20 light years. Once again, we are looking not for any old rough match to Earth, but one that has near-Earth-like gravity. Its atmosphere must be balanced enough that its carbon dioxide does not incinerate the planet in a Venus-like greenhouse, and that is breathable. That means it must have a biosphere to generate free oxygen. That also means that the immune system-compromised travelers will have to fend off entirely new diseases with immune systems that can barely fend off the common cold.

These are, collectively, not incidental challenges but are dictated by both astronomical and biological necessity. They have almost nothing to do with the technology we will use, which is the common way to discuss interstellar travel. We cannot easily solve them the way we might solve some types of technological problems because they are fundamental to our very make-up as humans. Evolution tuned our genome to survive comfortably on Earth with its unique physical makeup and biosphere. To thrive on another planet we will have to do more than just patch-up our current physiology with 'wonder pills' that compensate for gravity, atmosphere and alien pathogens. We can either do this at our destination by permanently living under a dome and only exploring our new world in spacesuits or robotic probes, or we will have to re-engineer our genome.

Epilog

The good news is that our search for a destination that warrants the multi-trillion-dollar effort and sacrifice by a human crew is now intimately bonded to the search for distant planets that have definite signs of biospheres. It is exciting to know that on the day we make the discovery of life existing on a remote planet beyond our solar system, we will at last have a destination that justifies the expense of a direct visit…but not before then.

Meanwhile, there is much we can do here at home that will fully engage us. Interplanetary travel is already a reality for our robotic emissaries. It is inevitable that human explorers will soon follow, perhaps in only a few decades. With an entire solar system available to us, advances in rapid travel and rocket technology will make even distant Pluto as accessible as modern-day Antarctica. For the next few centuries we have the opportunity to turn many corners of our solar system into thriving human colonies, mining local resources, and learning how to live and thrive in space.

Let's get busy and turn our solar system into a veritable suburb of New York City!

Interstellar Travel

Image Credits

History of Interstellar Travel
4-- Book cover image 1928 by Willey Ley in German.
10-- The Young Adventurer's Pocket Book of Space Travel, 1951)
13--A selection of rockets from the 1950s:http://tinyurl.com/panuyba
32--Dauntless ship: http://tinyurl.com/mj4yykb
24--Woodcut by Flammarion - Wikipedia
51--Standard Model table – Author
52--Quantum Corrall IBM: http://tinyurl.com/l5qjumj
55--Calabi-Yau spaces; Credit Jeff Bryant. http://tinyurl.com/o62q5tz

Where would we go?
62--Transit of Venus:Solar Dynamics Observatory in 2012: NASA/SDO
63--Exoplanet table: http://tinyurl.com/lmtnjuu

Habitable Zones
69--Habitable zone http://tinyurl.com/k3rsuzg

List of Destinations…so far!
77--HD 69830   Wikipedia/NASA JPL/T.Pyl
81--Fomalhaunt planet NASA/Hubble
82-- HR8799 – Wikipedia/Ben Zuckerman
80-- Data table: Chester Harman/PHL/NASA/JPL

Flare Stars
89--Very young star:NASA/Casey Reed
92—EPIC planetary system HEC: Graphical Catalog Results - Planetary Habitability Laboratory   @ UPR Arecibo

Are we there yet?
97--Barnard's Star motion. http://tinyurl.com/n4wcrzq
95-- Doppler shift http://tinyurl.com/lh239yp

Exoplanet Atmospheres
106--Pressure diagram: Karl Tater/LifesLittleMysteries.com/NASA
108--Infrared transit spectrum: NASA/Spitzer.
113--Habitable zone atmosphere model http://tinyurl.com/lrmyq9c

A Matter of Extreme Gravity
115--Human acceleration tolerance: NASAhttp://tinyurl.com/la7nkzb

Are we missing something?
123--Mass vs distance for exoplanets and detection methods – Author

Planetary Moons

Image Credits

128--Huge planet rising art: WallpaperHi.com
Space is not at all empty!
130--The Student Dust Counter (SDC), NASA/New Horizons
131--Interplanetary dust particle: Donald E. Brownlee. Wikipedia
137--Local chimney diagram. U.C. Berkeley http://tinyurl.com/
141--Debris frequency figure: http://tinyurl.com/k8xy34a
142--Half-inch hole Space Shuttle Endeavour : NASA
Radiation
145--Hand x-ray image: http://www.sunsetradiology.net/xrays.html)
149--Radiation dosage pie graph: World Nuclear
151--Cosmic ray spectrum. http://tinyurl.com/mghcdl9
153--Active shielding study. NASA: http://tinyurl.com/m6r9u6
Mutations
164--Population size and mutations. Cameron Smith.
165--Bar graph of radiation exposure: NASA/JPL-Caltech/SwRI
Diseases
171--Streptococcus bacteria found on the Surveyor spacecraft: NASA
178--Immune system pathways – http://tinyurl.com/p83oy3e
Physiological Effects
180--Image of astronaut's eye: Radiological Society of North America.
181--2001:A space odyssey – space station
184--ONeill Habitat – Wikipedia/ Rick Guidice, NASA
Psychiatric Effects
186--Inside the ISS. NASA
191--Jones the cat - Brandywine Productions.
192--Earth and moon from Cassini: NASA/Cassini
194--Bruce McCandless drifting above Earth: NASA
197--Astronaut quarters in spaceship: 2001: A space odyssey
Us versus Them
200--Pie graph religions: Wikipedia
203--Astronauts celebrating Christmas on the ISS in 2009: NASA
Because it's the Law!
206--Per capita crime by country: http://tinyurl.com/l49vcwy\
211--Space Leaders game - http://tinyurl.com/qedejl3
What's for dinner?
217--Mars Homestead Colony hydroponic garden - MarsHome.org
219--3-D Candy – HitechHalos
Diurnal cycles
225--Mars sunrise: NASA
227--Mars rover control – NASA/JPL

Interstellar Travel

Communication
233--Antenna pattern: MicrowaveSoft. http://tinyurl.com/k7ypwyo
235--NRAO 100-meter telescope: NRAO
237--The SBX-1 radar: US DoDefense. http://tinyurl.com/mdesdxk
238--Apache Point Laser: Image credit:Unknown
240--The digital camera image: Pixpolar.
243--National Ignition Facility: Damien Jemison/LLNL

Things that Break
247--MicroATX motherboard: Wikipedia/ Jonathan Zander
249--The AN/FSQ-7 Combat Direction Central computer: Wikipedia
251--Bendix G-15 computer
252--NASA printed rocket part: NASA /MSFC /Emmett Given
253--Printing microstructures: MIT Tech Review
254--Scanning Tunneling Microscopy, iron atoms: IBM Research

Its cold in space!
257--Blackbody curves for infrared: http://tinyurl.com/lfnjxec
258--House in infrared - http://tinyurl.com/bdpm87p
260--US Energy diagram: http://tinyurl.com/nzktwaa

Fuel and energy
264--Sketch of photocell: PVEducation.org
265--Spectrum of sun;solar Panasonic, http://tinyurl.com/mhhko28

Free Rocket Fuel
268--Ulysses spacecraft interstellar dust: http://tinyurl.com/l6wxbpf
269--The Bussard interstellar spacecraft: Wikipedia/NASA
273--Bubble chamber image: Brookhaven National Labs)
275--Magnetic bottle: http://tinyurl.com/mzbx3rt
278--Large Hadron Collider in CERN: CERN/LHC
279--Casimir force graph http://tinyurl.com/owbg42l

Down and Back
283--Animals at a waterhole in Africa: Google Earth.
288--X-43A Hypersonic Experimental Vehicle: Wikipedia/NASA

How much will it cost?
296--The Daedalus starship design / Saturn V. Credit: Adrian Mann.

The Miracle Cure: Stasis?
302--Artwork of hibernating astronauts: NASA
306--Astronauts sleeping in 2001:A space odyssey

AI and VR
313--Exponential computer growth: Wikipedia/Ray Kurzweil.
314--Head-mounted display: Wikipedia/NASA.
317--Brain chip: DailyTech/Institute of Neuro Informatics

Image Credits

Interstellar Rocket Technology
330--Project Orion: Wikipedia/NASA
332--IKAROS Solar Sail: Wikipedia/Andrzej Mirecki
'Warp factor 4 Mr. Sulu!'
338--This figure shows what this would look like:NASA
The Optimist's View
361--Space art by Don Davis under contract to NASA.
Conversations with the Public
374---NASA's warp ship design. http://tinyurl.com/lva2e8o

Bibliography

History of Interstellar Travel
---A great website for deckplans for science fiction-based rockets. http://www.projectrho.com/public_html/rocket/deckplans.ph
---An artist who has re-created rockets from the early-1900s. https://www.pinterest.com/charlwrites/aeroplanes-rockets-and-bikes-by-ollie/)
---The famous 'Lensman' and 'Skylark' box art. http://www.fantastic-plastic.com/BRITANNIA%20II%20PAGE.htm)
---The Dauntless, http://tinyurl.com/mj4yykb
---Parallel universes image rendered by an artist. http://www.unariuswisdom.com/parallel-universes-do-exist/
---The many-worlds idea in physics - http://listverse.com/2013/02/22/10-mind-bending-implications-of-the-many-worlds-theory/
---An extensive collection of real images of atoms re-arranged with technology by IBM http://www.international.ucla.edu/asia/article/9580
---A rendering of the 'hidden dimensions' of Calabi-yau geometry http://aether.lbl.gov/bccp/dimensions.html

Where would we go?
--- Habitable planets- http://phl.upr.edu/projects/habitable-exoplanets-catalog/media/pte
---Current tally of the different classes of exoplanets. http://exoplanetarchive.ipac.caltech.edu/docs/counts_detail.html
http://exoplanets.org/table
---Exoplanet classifications based on size, temperature and location http://phl.upr.edu/library/notes/athermalplanetaryhabitabilityclassificationforexoplanets
--- List of habitable planets http://en.wikipedia.org/wiki/List_of_potentially_habitable_exoplanets

Bibliography

---What can GAIS proper motions tell us about milky way dwarf galaxies?, Shoko Jin, Amina Helmi and Maarten Breddelsy
http://arxiv.org/pdf/1502.01215.pdf
---Proper motion across the sky of Barnard's Star
http://oneminuteastronomer.com/8869/barnards-star/

The list of destinations so far
---Alpha Centauri planet study http://arxiv.org/pdf/1208.2273v2.pdf

Exoplanet atmospheres
---Exoplanetary Atmospheres, Nikku Madhusudhan, Heather Knutson, Jonathan J. Fortney, Travis Barman
http://arxiv.org/pdf/1402.1169v1.pdf
---NASA's TESS mission cleared for next development phase, MIT News Office , November 10, 2014
http://newsoffice.mit.edu/2014/nasa-tess-exoplanets-mission-cleared-next-development-phase-1110

Are we missing something?
---Transit probabilities for stars with stellar inclination constraints, Thomas G. Beatty and Sara Seager 2010 ApJ 712 1433
http://iopscience.iop.org/0004-637X/712/2/1433/article

Space is not at all empty!
---The student dust counter on the new horizons mission
M. Horanyi, et al.,
http://www.boulder.swri.edu/pkb/ssr/ssr-sdc.pdf
---Dust measurements by the student dust counter onboard the new horizons mission to Pluto, 2015 , J. Szalay, M. Piquette, M. Horanyi,
http://www.hou.usra.edu/meetings/lpsc2015/pdf/1701.pdf
---This is a scanning electron microscope image of an interplanetary dust particle Donald E. Brownlee, University of Washington, Seattle, and Elmar Jessberger, Institut für Planetologie, Münster, Germany.
---Interstellar dust in the solar system, Harald Kruger, et al. 2007,
http://arxiv.org/pdf/0706.3110v1.pdf

Interstellar Travel

---Micron-sized dust particles detected in the outer solar system by the voyager 1 and 2 plasma wave instruments D. A. Gurnett, J. A. Ansher, W. S. Kurth, and L. J. Granroth, 1997.
http://onlinelibrary.wiley.com/doi/10.1029/97GL03228/pdf
---Viewing the Zodiacal Light, Kirit Karkare, 2012, http://astrobites.org/2012/05/21/viewing-the-zodiacal-light/
---The Local Chimney figure-
http://www.miqel.com/space_photos_maps/galactic_info/galactic-chimney-around-sun.html
---Orbital debris radar measurements, NASA orbital debris program office, http://orbitaldebris.jsc.nasa.gov/measure/radar.html
---ESA radar detects first orbital debris, 2014,
http://www.esa.int/Our_Activities/Operations/Space_Situational_Awareness/ESA_debris_radar_detects_first_space_objects
---Asteroid radar research, Steven Ostro,
http://echo.jpl.nasa.gov/introduction.html and
http://echo.jpl.nasa.gov/asteroids/goldstone_asteroid_schedule.html
---A Guide to Orbital Space Debris, 2010, Australian Space Academy, http://www.spaceacademy.net.au/watch/debris/gsd/gsd.htm

Communication
---A 10 kW 1060-nm-emitting YLS-10000-Y13 fiber laser from IPG Photonics (Oxford, MA)
http://www.laserfocusworld.com/articles/2014/10/nist-launches-new-calibration-service-for-high-power-lasers-up-to-10-kw.html
---Modified Internal Gate MIG Technology in Security and Surveillance Applications, 2015, PixPolar, Inc.
http://www.pixpolar.com/2012/09/modified-internal-gate-mig-technology-in-security-and-surveillance-applications/
---How many photons does the sun emit in a narrow band? Use this calculator to find out.
http://www.spectralcalc.com/blackbody_calculator/blackbody.php
---National Ignition Facility makes history record 500 terawatt shot
https://www.llnl.gov/news/national-ignition-facility-makes-history-record-500-terawatt-shot

Bibliography

---National Ignition Facility, Lawrence Livermore National Laboratory. http://tinyurl.com/kphk6ly

Atmosphere and Gravity
---Environment of Manned Systems, E.F.Adolph et al, 1958, http://history.nasa.gov/conghand/mannedev.htm

Because it's the Law!
---*Leadership issues with multicultural crews on the international space station: lessons learned from Shuttle/Mir.*, 2005, Kanas N1, Ritsher J., Acta Astronaut. 2005 May-Jun;56(9-12):932-6, http://www.ncbi.nlm.nih.gov/pubmed/15835051
---*The Code of Conduct for International Space Station Crews*, A. Farand, 2001, ESA bulletin 105 — february 2001, http://www.esa.int/esapub/bulletin/bullet105/bul105_6.pdf
---http://www.moc.noaa.gov/shipjobs/
---*Nature vs nurture: are leaders born or made? A behavior genetic investigation of leadership style*, 1998, Johnson AM1, Vernon PA, McCarthy JM, Molson M, Harris JA, Jang KL., Twin Res. 1998 Dec;1(4):216-23. http://www.ncbi.nlm.nih.gov/pubmed/10100814
---*Leadership and Brain Science: Possibilities for Harnessing Social Cognitive Neuroscience to Assess, Understand, and Develop Inspirational Leaders*, 2012, Dan DeFoe, http://tinyurl.com/kesn9gw
---*How Does the Psychology of Brain Function Affect Leadership Skills?*, 2012, Kirk Hallowell
http://www.aboutleaders.com/how-does-the-psychology-of-brain-function-affect-leadership-skills/

What's for dinner?
---U.S. could feed 800 million people with grain that livestock eat, Cornell ecologist advises animal scientists, 1997, Cornell Chronicle, http://www.news.cornell.edu/stories/1997/08/us-could-feed-800-million-people-grain-livestock-eat
---Rethinking the Meat Guzzler, 2008, Mark Bittman,

Interstellar Travel

http://www.nytimes.com/2008/01/27/weekinreview/27bittman.html?pagewanted=all&_r=0
---Chocolate linked to weaker bones, 2008,
http://articles.mercola.com/sites/articles/archive/2008/02/02/chocolate-linked-to-weaker-bones-but-what-kind-of-chocolate.aspx
---New studies of greenhouses connected to human quarters suggests that enough plant oxygen would be generated in a few weeks to become a fire hazard.
---Farming on Mars: NASA Ponders Food Supply for 2030s Mission, 2013, Clara Moskowitz
http://www.space.com/21028-mars-farming-nasa-missions.html
---Mars colonization may require Earth soil, 2013,
http://www.slate.com/articles/technology/future_tense/2013/06/mars_colonization_may_require_earth_soil.html
---A Vegan colony on Mars, Evolving the human diet, 2015, BlissfulEating.com,
http://www.blissfuleating.com/vegan-colony-mars-evolving-human-diet/
---3D printed pizza to feed Mars colonists?, 2013, Ian O'Neil,
http://news.discovery.com/space/3-d-printed-pizza-to-feed-mars-colonists.htm
---Food you can print, and it is all thanks to 3D printing, 2015,
https://hitechhalos.wordpress.com/2015/02/23/printed-food

Radiation
---What is radiation?, World Nuclear Organization, http://www.world-nuclear.org/Nuclear-Basics/What-is-radiation-/
---The Geology of Radon in Kansas, 2007, Shane Lyle,
http://www.kgs.ku.edu/Publications/PIC/pic25.html
---Cosmic ray spectra, Whanlon,
http://www.physics.utah.edu/~whanlon/spectrum.html
---Technical aspects of the air scrubber machines, Scrubber.com,
http://www.skyscrubber.com/3000_machines.htm
---How much radiation will the settlers be exposed to?, 2015, MarsOne,

Bibliography

http://www.mars-one.com/faq/health-and-ethics/how-much-radiation-will-the-settlers-be-exposed-to and,
http://www.marsjournal.org/contents/2006/0004/files/rapp_mars_2006_0004.pdf
---Rapid Diagnosis in Populations at Risk from Radiation and Chemicals, 2010, edited by Antonina Cebulska-Wasilewska, Andreyan N. Osipov, Firouz Darroudi, http://tinyurl.com/n75lzjh
---Health effects of nuclear radiation in plain language, 2015, Jerry Cuttler,
http://atomicinsights.com/health-effects-nuclear-radiation-plain-language/
---Radiation hazard of relativistic interstellar flight, O. Semyonov http://arxiv.org/ftp/physics/papers/0610/0610030.pdf

Magnetic fields as shielding
---Radiation hazard of relativistic interstellar flight, O. Semyonov http://arxiv.org/ftp/physics/papers/0610/0610030.pdf
--Shielding.
https://engineering.dartmouth.edu/~d76205x/research/Shielding/docs/vonBraun_69.pdf
---Magnetic Shielding for Spacecraft, 2005, N. Atkinson, http://www.thespacereview.com/article/308/1

Flare Stars
---A Dust Ring Around Epsilon Eridani: Analogue to the Young Solar System, 1998, Greaves, J., Holland, W., Moriarty-Schiven, G., Dent, W., Zuckerman, B., McCarthy, C., Webb, R., Butner, H., Gear, W., and Walker, H. in Astrophysical Journal Letters, 506, L133 (1998).

Mutations
---Want to colonize an alien planet? Send 40,000 people, 2014, Mike Wall,
http://www.space.com/26603-interstellar-starship-colony-population-size.html

Interstellar Travel

---Human voyages to Mars pose higher cancer risks, 2013, Ken Kremer,
http://www.universetoday.com/102487/human-voyages-to-mars-pose-higher-cancer-risks/
---Would a trip to Mars damage your brain?, 2011, Jim Schnabel, http://www.dana.org/News/Details.aspx?id=43154
---Mission to Mars could mess with your brain, 2013, Ian O'Neill, http://news.discovery.com/space/private-spaceflight/mission-to-mars-could-cause-brain-damage-130102.htm

Other Physiological Effects
---MRI Shows Some Astronauts Experience Brain and Eye Abnormalities, 2012, Marijke Durning,
 http://www.diagnosticimaging.com/mri/mri-shows-some-astronauts-experience-brain-and-eye-abnormalities
---Rotating Space Station Stabalization Effects, 1969, Carl Larson et al, http://tinyurl.com/kxso35u

Psychiatric Effects
---It's All about People: NASA Psychiatrist Explains Why Space Itself Is Not Detrimental, 2012, Tereza Pultarova,
http://www.spacesafetymagazine.com/spaceflight/commercial-spaceflight/its-people-nasas-shrink-explains-space-detrimental-human-mind/
---Risk of Behavioral and Psychiatric Conditions, Slack, K. et al., http://tinyurl.com/k63fcg6
---What does space travel do to your mind? NASA's resident psychiatrist reveals all., 2012, Esther Inglis-Arkell,
http://io9.com/5967408/what-does-space-travel-do-to-your-mind-nasas-resident-psychiatrist-reveals-all
---Psychosocial issues in long-term space flight: overview, L. A. Palinkas, http://web.mit.edu/16.459/www/Palinkas.pdf

Us versus Them
---Survey on beliefs. http://tinyurl.com/lcc7wn3

Bibliography

---Study eyes influence of religion on future space exploration, 2014, Leonard David, http://www.space.com/27896-religion-influence-space-exploration.html

---Muslim clerics issue fatwa banning the devout from Mars One 'suicide' mission, 2014, Iain Thomson, http://www.theregister.co.uk/2014/02/22/muslim_clerics_issue_fatwa_banning_the_devout_from_mars_one_suicide_mission/

---Space and religion: how believers view latest space developments, 2013, Herb Scribner, http://tinyurl.com/qa2fenk

---Creationist calls for end to space exploration because aliens go to hell, 2014, Michael Stone, http://tinyurl.com/k3udbq7

---Should humanity take religion on interstellar space voyage?, 2012, Clara Moskowitz, http://www.space.com/17659-interstellar-travel-religion-conflict.html

Funny odors

---How NASA Deals With Odor Inside the International Space Station, 2014, Robert Frost and Clayton C. Anderson http://gizmodo.com/how-nasa-deals-with-odor-inside-the-international-space-1648864449

---The Smell of Space, Expedition Six, Space Chronicles #4, 2003,: ISS Science Officer Don Pettithttp://spaceflight.nasa.gov/station/crew/exp6/spacechronicles4.html

---Astronauts in space lose their sense of smell and crave Tabasco sauce!, 2014, Jose Duarte, http://www.omgfacts.com/lists/11467/Astronauts-in-space-lose-their-sense-of-smell-and-crave-Tabasco-sauce-abf5

---Astronaut Don Pettit Diary: http://spaceflight.nasa.gov/station/crew/exp6/spacechronicles4.html

--- https://spinoff.nasa.gov/Spinoff2013/cg_4.html

Interstellar Travel

Diurnal cycles
---Sleep and circadian rhythms in four orbiting astronauts, 1998, Monk TH1, Buysse DJ, Billy BD, Kennedy KS, Willrich LM., J Biol Rhythms. 1998 Jun;13(3):188-201
http://www.ncbi.nlm.nih.gov/pubmed/9615283
---Wide awake in outer space, 2001, NASA Science News,
http://science.nasa.gov/science-news/science-at-nasa/2001/ast04sep_1/
---Keeping the right time in space: importance of circadian clock and sleep for physiology and performance of astronauts, 2014,
Jin-Hu Guo, et al., Military Medical Research 2014, 1:23.
http://www.mmrjournal.org/content/pdf/2054-9369-1-23.pdf
---Step into the Twilight Zone: Can Earthlings Adjust to a Longer Day on Mars?, 2013, Katie Worth,
http://www.scientificamerican.com/article/step-into-the-twilight-zone-can-earthlings-adjust-to-a-longer-day-on-mars/

The Miracle Cure: Stasis?
---How Astronaut Hibernation for Deep-Space Travel Works, 2013, Karl Tate,
http://www.space.com/22526-astronaut-hibernation-space-travel-infographic.html
---Is human HIBERNATION the key to getting to Mars? Putting astronauts into a 'coma' could make reaching the red planet easier and cheaper, 2014, Jonathan O'Callaghan, http://tinyurl.com/kux6796

Therapeutic Hypothermia for Brain Ischemia
---Where Have We Come and Where Do We Go?, 2010, Midori A. Yenari, MD; Thomas M. Hemmen, MD, PhD
http://stroke.ahajournals.org/content/41/10_suppl_1/S72.full
---Clinical applications of induced hypothermia, Mark Luscombe, John C Andrzejowski,
http://ceaccp.oxfordjournals.org/content/6/1/23.full
---Therapeutic hypothermia in patients with aneurysmal subarachnoid hemorrhage, refractory intracranial hypertension, or cerebral

Bibliography

vasospasm. 2009, Seule MA1, Muroi C, Mink S, Yonekawa Y, Keller E., Neurosurgery. 2009 Jan;64(1):86-92; discussion 92-3.
http://www.ncbi.nlm.nih.gov/pubmed/19050656

Artificial Coma.
---What Is a Medically Induced Coma and Why Is It Used?, 2011 David Biello,
http://www.scientificamerican.com/article/what-is-a-medically-induced-coma/
---What Is a Medically Induced Coma?, 2013, Marc Lallanilla,
http://www.livescience.com/39483-what-is-a-medically-induced-coma.html
---Joan Rivers Out Of Coma, On Life Support: What's A Medically Induced Coma, Anyway?, 2014, Samantha Olson
http://www.medicaldaily.com/joan-rivers-out-coma-life-support-whats-medically-induced-coma-anyway-300836
---Human hibernation: Secrets behind the big sleep, 2014, Frank Swain, http://tinyurl.com/lu6pw3d

Diseases
---Hunting for Disease Genes, 2011, Nicholas Wright Gillham
http://www.ftpress.com/articles/article.aspx?p=1692537&seqNum=4
---British science targets nearly half of world's genetic diseases, 2001, Wellcom Trust Sanger Institute,
http://www.sanger.ac.uk/about/press/2001/publication2001/britishrole.html
---Definition of Genetic Disease, Medicine.net,
http://www.medicinenet.com/script/main/art.asp?articlekey=31302
---The Growth of Bacterial Populations,2012, Kenneth Todar
http://www.textbookofbacteriology.net/growth_3.html
---Is Dirt Good for Kids? Are parents keeping things too clean for their kids' good?, Lisa Zamosky
---Study Reveals Immune System is Dazed and Confused During Spaceflight, 2014, NASA, http://tinyurl.com/9n7m78p

Interstellar Travel

http://www.nasa.gov/content/study-reveals-immune-system-is-dazed-and-confused-during-spaceflight-u
---More Evidence that Space Travel Is Bad for Immune Systems, 2014,Michael Byrne http://tinyurl.com/l6ar9zh
---Weightless space travel may suppress immune system,Kelly Young, http://tinyurl.com/qjcw27c

Boredom
---The Long Journey Through Nothingness, http://mysimplereality.com/?page_id=3916
---How will humans cope with boredom on their first missions to Mars?, 2013, Robbie Gonzalez, http://tinyurl.com/plggkue
---Space Boredom Busters: Movies, Music, Jokes, 2007, Gina Sunseri, http://abcnews.go.com/Technology/story?id=3777490&page=2
---Bordom in Antarctica - http://tinyurl.com/lb2gtxb

Things that Break
---How many parts is each car made of?, Toyota, http://www.toyota.co.jp/en/kids/faq/d/01/04/
---Saturn V is the Biggest Engine Ever Built, 2004, http://www.popularmechanics.com/science/a227/1280801/
---Space Shuttle Era Facts, NASA, http://tinyurl.com/4xf8h3a
---NASA 3-D Prints First Full-Scale Copper Rocket Engine Part, 2015, NASA, http://tinyurl.com/ntcrczl
---BioBots Wants to Be Your Desktop Rapid Bio-Prototyping Fabrication Station, 2015, J.F. Brandon, http://tinyurl.com/ne9f944
---Micro 3-D Printer Creates Tiny Structures in Seconds, 2013, Prachi Patel, http://tinyurl.com/aeopm8x
---STM Image Gallery, IBM Watson Labs, http://tinyurl.com/qc3yxgk

Fuel and energy
---First Photovoltaic Devices, PVEducation.org, http://www.pveducation.org/pvcdrom/manufacturing/first-photovoltaic-devices

Bibliography

---High-performance flat-panel solar thermoelectric generators with high thermal concentration, 2010, Daniel Kraemer et al., Nature Materials 10, 532–538 (2011)
http://www.nature.com/nmat/journal/v10/n7/full/nmat3013.html
---A Level Physics Notes: Electricity – The Magnetic Bottle,
http://astarmathsandphysics.com/a-level-physics-notes/electricity/a-level-physics-notes-the-magnetic-bottle.html
---Taking a closer look at LHC, Xabier Cid Vidal & Ramon Cid
http://www.lhc-closer.es/1/3/9/0

Down and Back
---Hypersonic Jets That Zoom Into Orbit, 2001, Michael Cabbage
http://articles.orlandosentinel.com/2001-04-15/news/0104150338_1_nasa-rocket-scientists-research
---Hypersonic Flight,
http://www.geocities.ws/spacetransport/hypersonic.html

How much will it cost?
---State of the Satellite Industry Report, 2014, SIA.org,
http://www.sia.org/wp-content/uploads/2014/09/SSIR-September-2014-Update.pdf
---Global Wealth Report, 2014, http://tinyurl.com/pqar9wn
---World Military Spending, 2013, Anup Shah
http://www.globalissues.org/article/75/world-military-spending
---A Mars Mission for Budget Travelersz; Twenty years and $100 billion could get us there, panel says., 2014, Marc Kaufman,
 http://news.nationalgeographic.com/news/2014/04/140422-mars-mission-manned-cost-science-space/
---Mars for Only $1.5 Trillion, 2015, O. Glenn Smith and Paul D. Spudis, http://spacenews.com/op-ed-mars-for-only-1-5-trillion/
---"Evaluation of Technological/Social and Political Projections for the Next 300 Years and Implications for an Interstellar Mission," Journal of the British Interplanetary Society Vol. 65 (2012), pp. 330-340, http://www.centauri-dreams.org/?p=31774

Interstellar Travel

---Three Scenarios For Funding Interstellar Travel, 2013, Elise Ackerman, http://tinyurl.com/ov53aqb
---Interstellar Space Flight: Social and Economic Considerations, 2010, Richard Obousy, http://www.icarusinterstellar.org/interstellar-space-flight-social-economic-considerations/
---Machine Intelligence, the Cost of Interstellar Travel, and Fermi's Paradox, 1994, Louis K. Scheer, http://www.lscheffer.com/qjras.pdf
---Neuron Chip "Learns" to Recognize Distinct Gestures Via Retina Sensors, 2013, Jason Mick http://tinyurl.com/nztncw5
---Interstellar travel cost http://tinyurl.com/4fg4vxq

Energy costs
---We Won't Have Enough Power For Interstellar Travel Until At Least 2211, According to New Calculations, 2011, Rebecca Boyle, http://www.popsci.com/science/article/2011-01/interstellar-travel-wont-be-possible-least-200-years-according-new-calculations
---The 100-year Starship Program, http://100yss.org/

AI and VR
---Population Estimates: Year One through 2050 A.D., http://www.ecology.com/population-estimates-year-2050/
---Public Opinion Polls and Perceptions of US Human Spaceflight, 2003, Roger Launius, Space Policy 19 (2003) 163–175 http://www.academia.edu/179045/_Public_Opinion_Polls_and_Perceptions_of_US_Human_Spaceflight_
---NASA popularity still sky-high, 2015, Seth Motel, Pew Research Center, http://tinyurl.com/olffdpt
---Computers and the Internet history, http://www.futuretimeline.net/subject/computers-internet.htm
---Computers to match human brains by 2030, 2015, http://www.independent.co.uk/life-style/gadgets-and-tech/news/computers-to-match-human-brains-by-2030-782978.html
---Immersive telepresence as a new paradigm for mars exploration, 2012, R. J. Terrile,

Bibliography

http://www.lpi.usra.edu/meetings/marsconcepts2012/pdf/4169.pdf

Interstellar Rocket Technology
---World record for compact 'tabletop' particle accelerator, 2014, Physics.org,
http://phys.org/news/2014-12-world-compact-tabletop-particle.html
---Proton and Carbon Ion Therapy, 2013, edited by C-M Charlie Ma, Tony Lomax, Taylor and Francis Group publishing, Page 210.,
http://tinyurl.com/qzmjtyr
---Proton Beam Radiotherapy - The State of the Art, 2005, Harald Paganetti and Thomas Bortfeld, in New Technologies in Radiation Oncology (Medical Radiology Series) (Eds.) W. Schlegel, T. Bortfeld and A.-L. Grosu, Springer Verlag, Heidelberg, http://www.aapm.org/meetings/05AM/pdf/18-4016-65735-22.pdf
---Design of a compact synchrotron for medical applications, 2003, Nader Al Harbi and S. Y. Lee, Review of scientific instruments, volume 74, number 4
https://uspas.fnal.gov/programs/masters-degree/Al-HarbiThesis.pdf
---Rocket engine exhaust velocity calculation,
https://www.physicsforums.com/threads/rocket-engine-exhaust-velocity.296023/
---First Interstellar Spacecraft May Use Texas-Size Solar Sail, 2013, Mike Wall, http://www.space.com/20169-interstellar-spaceflight-solar-sail.html
---Laser Beamed Interstellar Mission: A New Take, 2008, Paul Gilster, http://www.centauri-dreams.org/?p=3493
---Incredible Technology: How Solar Sails Could Propel the First Starships, 2013, Miriam Kramer,
http://www.space.com/22442-solar-sail-starship-interstellar-spacecraft.html
---Focus on the Sail, 2013, Paul Gilster
http://www.centauri-dreams.org/?p=28274
---Roundtrip Interstellar Travel Using Laser-Pushed Lightsails, 1984, Robert L. Forward, Journal of Spacecraft, vol. 21, no. 2, march-April, p. 184., http://www.lunarsail.com/LightSail/rit-1.pdf

Interstellar Travel

Warp factor 4 Mr. Sulu!
---The warp drive: hyper-fast travel within general relativity, 1994, Miguel Alcubierre, Class. Quantum Grav. 11 L73, http://iopscience.iop.org/0264-9381/11/5/001
---This is NASA's new concept spaceship for warp drive interstellar travel, 2014, Gizmodo.com, http://tinyurl.com/mel84oh
---Galactic Matter and Interstellar Flight, 1960, Robert Bussard, Astronautica Acta 6 (4): 179–194. http://tinyurl.com/kvojw97
---Magnetic sails and interstellar travel, 1990, Dana Andrews and Robert Zubrin, Journal of The British Interplanetary Society, Vol 43, pp. 265-272,1990

Fermi's Paradox Reconsidered
---Where is Everybody? An Account of Fermi's Question, 1985, Jones, E. M., Department of Energy Research Accomplishments, http://tinyurl.com/m8bf9xd
---Where is ET? Fermi's Paradox Turns 65, 2015, David Bailey and Jonathan Borwein, http://tinyurl.com/p478u7t
---SETI @ Home, website, http://setiathome.ssl.berkeley.edu/
---Project Phoenix, http://www.seti.org/seti-institute/project/details/project-phoenix

It's Cold in Space
---Cloth blocks infrared http://tinyurl.com/l7we5yu

Made in the USA
Lexington, KY
11 June 2015